Girolamo Cardano

1501–1576

Girolamo Cardano

1501–1576

Physician, Natural Philosopher, Mathematician, Astrologer, and Interpreter of Dreams

Markus Fierz

Translated by Helga Niman

1983
Birkhäuser
Boston • Basel • Stuttgart

Markus Fierz
Felseneggstrasse 10
CH-8700 Küsnacht
Switzerland

Edited for the Swiss Federal Institute of Technology, Zürich, and its Department of Humanities and Social Sciences by

Jean-François Bergier
Markus Fierz
Roger Kempf
Adolf Muschg
Hans-Werner Tobler

Library of Congress Cataloging in Publication Data
Fierz, Markus.
Girolamo Cardano (1501–1576)
Translation of: Girolamo Cardano (1501–1576)
1. Cardano, Girolamo, 1501–1576.
2. Physicians—Italy—Biography.
3. Mathematicians—Italy—Biography.
4. Astrologers—Italy—Biography. I. Title.
R520.C32F5313 610'.92'4 [B] 82-4173
AACR2

CIP–Kurztitelaufnahme der Deutschen Bibliothek
Fierz, Markus.
Girolamo Cardano: (1501–1576); physician, natural philosopher, mathematician, astrologer, and interpreter of dreams/Markus Fierz. Transl. from the German by Helga Niman.–Boston; Basel; Stuttgart: Birkhäuser, 1982.
Dt. Ausg. unter demselben Titel
ISBN 3-7643-3057-0

ISBN 3-7643-3057-0
Printed in USA

AVTORIS CARMEN.

Non me terra teget, cœlo ſed raptus in alto
Illuſtris uiuam docta per ora uirûm.
Quicquid uenturis ſpectabit Phœbus in annis,
Cardanos noſcet, nomen & uſq; meum.

Girolamo Cardano
1501–1576

THE PROFILE OF CARDANO reproduced on the preceding page appears for the first time on the title page of the Basel edition 1554 of *De Subtilitate* (Ludovicus Lucius anno 1554). A similar portrait, somewhat smaller, can be found on the title page of Cardano's Commentaries on Cl. Ptolemy's "Quadripartitum" (Basel, Henri Petri 1554). Both woodcuts are reproductions of a medal from the Mint of Milan ascribed to Leone Leone: vide G. F. Hill and G. Pollard, "Renaissance Medals from the S. H. Kress Collection," N. 436 (London, 1967).

All later portraits of Cardano go back to the said woodcuts. The well-known engraving in Vol. I of the "Opera" (1663) has the features accentuated to the demonic, reflecting the baroque, seventeenth century view of Cardano's personality.

Contents

	Preface	ix
	Preface to the English Edition	ix
	Introduction	xi
1	Cardano's Life and Writings	1
2	Cardano the Physician	37
3	Natural Philosophy and Theology	56
4	*De Subtilitate* and *De Rerum Varietate*	88
5	Astrology	117
6	The Interpretation of Dreams	125
7	On the Art of Living with Oneself	156
	Postscript	167
	Notes	177
	References	193
	Appendix	197

Preface

THIS STUDY OF GIROLAMO CARDANO is the work of an amateur in the field of the history of science and the history of ideas. As a mathematical physicist I lack the depth of training in philological-historical disciplines necessary to discuss the sources of Cardano's knowledge and trace the influences that shaped his views on science, medicine, and philosophy.

What little recent literature on Cardano there is sometimes shows a lack of true understanding, or is primarily an appreciation of the mathematician. I relied, therefore, largely on his own writings, which are collected in the *Opera Omnia.* My excerpts and translations are taken directly from these works, and I hope that I have succeeded in capturing their essential meaning and spirit.

MARKUS FIERZ

Preface to the English Edition

I THANK the publisher Birkhäuser Boston, Incorporated, for the venture of this English edition of my essay on Cardano that originally appeared in the Poly Series published by Birkhäuser Verlag, and for the care given to the translation.

For this edition I have added two longer sections: one on Cardano's voyage to Scotland, and another on a rather interesting "mathematical theosophy" contained in his *Liber de Proportionibus* (Basel, 1570). Also included is a list of references quoted in the notes, and a record—as complete as possible—of the original editions of Cardano's writings.

MARKUS FIERZ

Introduction

THIS *MONOGRAPH* is an attempt to acquaint the modern reader with the philosophy and scientific investigations of the universal scholar Girolamo Cardano of Milan, who lived from 1501 to 1576. Cardano was a physician, astrologer, and interpreter of dreams. His significance as a mathematician is undisputed; he enjoyed an international reputation as a physician among his contemporaries; and a hundred years later Leibniz still praised him as a man who was truly knowledgeable. He is the oldest of the famous Italian natural philosophers, and is usually mentioned together with Bernado Telesio (1508–1588) and Francesco Patrizzi (1529–1597). Cardano greatly influenced Patrizzi, and he surpasses both his colleagues in the wide range of his interests and knowledge, as well as in the power of his creative imagination. The latter does lead him occasionally toward the intellectually adventurous, and has prompted comparisons with his contemporary, Paracelsus (1493–1541).[1] Beside this volcanic visionary, however, Cardano seems dispassionate and systematic. Yet, when judged by present-day standards, his writings do appear rather fantastic and unmethodical, although they apparently did not make this impression on his contemporaries! Theirs was a time of great social upheaval and chaos. Throughout Cardano's youth and adulthood, the power struggles of Emperor Charles V (1500–1558), Francis I (1494–1547), and

Henry VIII (1491–1547) dominated the life of Europe and shaped its fate. A succession of wars engulfed Italy in turmoil and terror—that same Italy where Titian, Vignola and Palladio were creating their works.

The voluminous corpus of Cardano's writings[2] embraces all the major ideas of his time, and thus conveys an authentic picture of the intellectual life of the High Renaissance. His writings contain rudiments and ideas pertaining to almost every scientific and philosophical doctrine developed during the seventeenth century.

Yet, his name remained familiar only to mathematicians. As a philosopher and natural scientist he has practically been forgotten, in much the same way that the imaginatively comprehensive thought of the Renaissance has been generally forgotten. The seventeenth century strove above all for clarity and distinctness of conception and thought, a goal it finally reached in the scientific-mathematical method, which has asserted its predominance despite its extreme one-sidedness.

This "new science," as it developed toward the end of the sixteenth century, found its main antagonist in Scholasticism, that is to say, in the Aristotelian philosophy of nature taught in the universities. This prompted Galileo to introduce in his dialogues the figure of Simplicio, the Peripatetic opponent of the new doctrine, as representative of the philosophy that Galileo attacks. Simplicio is portrayed throughout as a learned and amiable gentleman who knows how to remain gracious even when attacked for expounding an obsolete theory. Renaissance philosophy, on the other hand, could not at that time be considered antiquated. It was anti-Aristotelian, relying on Plato, the Neo-Platonists and the fabled Hermes Trismegistus as its authorities. To Galileo, however, this philosophy and conception of nature was not a serious science, but rather, a form of medieval superstition. Yet these ideas were by no means medieval—they reflected the spirit of their time. Apparently, Galileo thought it beneath his dignity to discuss them seriously. He was, therefore, all the more surprised that Kepler, his colleague in the defense of the Copernican doctrine, considered so-called occult causes to be possible, and that he believed, for example, that the influence of the moon on the earth's bodies of water caused the tides.[3]

As a consequence of Galileo's attitude—or of that of Descartes, for that matter—the natural philosophy of the Renaissance has received little attention in the history of philosophy and science. The doctrines of Galileo's outspoken opponents were, on the other hand, carefully studied. Their merits and their shortcomings were evaluated, and it was concluded that during its second period Scholasticism made notable contributions to physics, on which Galileo himself relied.[4] These medieval investigations in the field of optics, mechanics, and the increase and decrease of quantities, and the scholars who carried them out—such as Vitelo, Jordanus, the Franciscans at Merton College, Oxford, the Parisian Scholastics like Buridan and Oresme—belong to the thirteenth and fourteenth centuries. A gap of more than two hundred years separates the High Middle Ages from the seventeenth century, and we hear little about the intellectual life of this extended period.

When judged by the standard of Galilean methods of investigation, the scientific philosophy of the Renaissance certainly is minor. But this disregards Kepler's importance and makes his views questionable or inexplicable. Galileo found them indeed incomprehensible, and Descartes preferred to ignore Kepler whenever possible, although in the field of optics he probably owed more to Kepler than he cared to admit. A modern author like Bertrand Russell[5] tries to view Kepler's achievements merely as the result of diligence, proving thereby that he does not have a proper understanding of the man nor a sense for his genius.

Still, it is indisputable that in many respects natural science in the fifteenth and sixteenth centuries is either unoriginal, or it shows a tendency toward the fantastic and the superstitious.

L. Thorndyke, in his *History of Magic and Experimental Science*, discusses in detail the literature of this period of transition, and he observes that its scientific level shows a steady and increasingly obvious decline.[6] In the first chapter of the fifth volume of his work he characterizes the state of the sciences in the first half of the sixteenth century. At this time, more than ever before, antiquity was regarded as the model while medieval science was rejected, if only on account of its language, which was considered barbaric. Furthermore, Scholasticism was viewed as pedantic, something a cultured person should never be.

The invention of the printing press encouraged the writing of popular works directed at a large audience. Works of this kind were also written in the Middle Ages, but they were rather good and concise basic textbooks. Now, works of an encyclopedic nature were written which aimed at engaging the reader's interest by the diversity of their frequently fantastic subject matter. The sixteenth century was not the age of the erudite specialist characteristic of the medieval scholar; rather, it was an age of the dilettante polyhistorian. Conrad Gessner, himself a polyhistorian, deplores the general lack of sound specialized knowledge. Everyone, he says, goes beyond the confines of his profession: the schoolmaster turns philosopher, the physician dabbles in astrology, the astrologer in medicine.

Ernst Cassirer[7] points out that in the sixteenth century,

> the concentration on the sensual wealth of the physical world and the endeavor to take hold of it directly and in all its aspects not only failed to establish the new specifically modern concept of nature, but indeed hampered its development. Until the medium of mathematics and new modes of thought produced by it established specific criteria of experience itself, the empiricism of the Renaissance lacked any objective standard or principle of selection in view of the wealth of data to be observed. In motley profusion but without any organizing principle the individual "facts" stand side by side. Reference to experience must remain a dubious foundation as long as its definition still includes totally heterogeneous elements. Natural philosophy of the fifteenth and sixteenth centuries provides the rudiments of exact description and exact experiment, but both these endeavours are paralleled by efforts to lay the foundation for an "empirical magic". . . . In this way, the empirical world not only borders on that of the miraculous, but both constantly merge into each other. The atmosphere of the "science" of nature is totally pervaded and saturated by notions of the miraculous.

Anyone even vaguely familiar with the scientific literature of the time will have to agree with this overall assessment. I would, however, question whether, as Cassirer believes, "the basis for exact experiment" was already laid at that time. Gilbert, Galileo, and their successors were the first to consciously practice experimental science. Certainly, earlier scholars readily pointed out that their assertions were supported by "manifold experiments." These "experiments" were, however, for the most part based on rather strange occurrences, which the author, moreover, usually only knew from hearsay. This cannot be regarded as experimental knowledge in the modern sense, knowledge derived from artificially created, controlled, and reproducible conditions.

Experience—"experientia"—that, as Cassirer puts it, borders on the realm of the supernatural, is not that of everyday life, easily accessible to everyone everywhere. Galileo was the first to make particular reference to such everyday experiences, and he recognized their potential as the basis of an exact science. Until that time, experiences deemed worthy of scientific investigation were primarily conspicuous and unusual phenomena that people did not understand and were, therefore, eager to explain. Cassirer also makes reference to the "chaotic irregularity" with which the "facts" are joined, thereby pointing out a tendency closely related to the rejection of the pedantry of Scholasticism that Thorndyke alludes to. As the scholastic mode of thought in its Aristotelian tradition was rejected, the appreciation of medieval scientific methods was lost as well.

Aristotelian philosophy had provided a general cosmological framework for every kind of research. In it, each phenomenon occupied the place which corresponded to its abstract essence; this accounted for the logical order of the universe. In view of this, the subtle logical distinctions did not seem in the least pedantic to the scholastics; they regarded them as essential to an understanding of the universal order. Since the universe was fashioned by a creator of great wisdom who also represented, or rather embodied, the supreme logical principle, scientific thought was not particularly interested in the peculiar and the miraculous. It moved quite objectively and without regard for worldly concerns. That the world is full of wonder was never denied; on the contrary, this was taken for granted and seen as continuous evidence of the divine creator. It was for this very reason that the investigation of a single problem was regarded as being as interesting as the compilation of astounding phenomena.

It is noteworthy that a similar decline of the scientific method can be observed in Hellenistic antiquity as the purport of the Aristotelian scientific method became less and less understood.[8] Then too, the Aristotelian method was considered far too rationalistic. Moreover, the idea of pure research had lost its appeal, whereas, to quote Festugière, "la science aristotélicienne est essentiellement contemplative: elle est une θεωρια. On vise à connaître pour connaître, non pas pour utiliser cette connaissance à des fins pratiques."

Pseudosciences with practical objectives now replace pure science. One studies the movement of the stars in order to divine human fate. One compiles the properties of animals, plants, even stones hoping to procure from them medicinal remedies. At first glance, then, it appears that science is disintegrating into the pursuit and collection of wondrous and mysterious phenomena. And yet, this way of looking at the world has its own peculiar intrinsic unity. The main principle underlying occult relationships is the sympathy and antipathy of all things which partake in a common life. This life is of celestial origin, streaming down onto the earth with the light of the stars and the warmth of the atmosphere. Closely related to these attempts to grasp the animation of all things and their sympathetic and antipathetic interactions is a religious motive, just as Aristotle's efforts to comprehend the logical-hierarchic order of the universe and its supreme, divine origin were motivated by religious concerns. It appears, however, that Aristotle's view of the world as a logical-hierarchic order could not, in the long run, satisfy these religious needs adequately. It must have seemed too academic to the average understanding. In the occult sciences the emotions try to assert their importance. Sympathy and antipathy are indeed emotions or principles of emotions, and they replace the logical opposites on which Aristotle's cosmology was based.

Throughout the Middle Ages the occult sciences, and with them the endeavours to understand the order of the universe intuitively and emotionally, accompanied Scholasticism like an undercurrent. During the fifteenth and sixteenth centuries, however, this current surfaces with renewed vigor, and undermines even the thinking of scholars of very rational disposition. One might compare these currents of intellectual history to a flooding of the Nile, which is destructive yet at the same time makes the ground more fertile. As the "waters receded," a new science sprang up in which new modes of thought came into being, and this science was not entirely dependent on the "medium of mathematics," as Cassirer says.

Cornelius Agrippa[9] (1486–1535) became the most eminent exponent of occult philosophy of this period, and despite his rather ambiguous personality, we have to consider him in our context. He called himself "von Nettesheim," insisting that he was of old aristocratic lineage.[10] He designated himself counsel to Emperor Charles V, and was generally given to recounting fabulous events of his life—bold military exploits, for example. Although he did not have a doctorate in medicine, he on occasion held professorships in medicine, and even served as personal physician to the mother of Francis I when she was Regent at Lyons during the King's Italian campaign. Agrippa lost this position when Francis was defeated in the battle of Pavia and was taken prisoner by the Emperor. In 1530, Agrippa had his two famous works, *De incertitudine et vanitate scientiarum declamatio invectiva* [On the vanities and uncertainties of the sciences] and *De occulta philosophia* [On occult philosophy and magic], published. In the more than one hundred (mostly very short) chapters of *De incertitudine et vanitate scientiarum* all kinds of human activities are presented as absurd and ridiculous. This is done with great verbosity, interspersed with quotations from numerous classical authorities, and it is not without a certain robust humor. It would be a misunderstanding of Agrippa's personality to take these utterances too seriously.[11] It was common practice among Renaissance authors to discredit and ridicule things commonly praised, or inversely, to praise that which most people found reprehensible. In this vein, things such as *In Praise of Syphilis* or *In Praise of Gout*[12] were written. Another book of this kind is *In Praise of Folly* by Erasmus, but it surpasses all others in its charm and subtle wit as it leaves judgment suspended, thus revealing how much more sincere Erasmus was than all the others. It is most likely that this book served Agrippa as a model, although he was unable to measure up to its standards. Nonetheless, *De vanitate scientiarum* is a characteristic product of its time and enjoyed great popularity. The same can be said of the book *De occulta philosophia*. Agrippa was probably the first to try to present these ideas comprehensively and systematically, to the extent that this is possible at all!

Central to these teachings is the idea of the sympathy of all things. In chapter thirty-seven of the first book, Agrippa expresses this as follows:

> The world is the image of God, man the image of the world, animals that of man, plant life that of the animals, metals that of plant life, and stones that of metals. With regard to spiritual matters, however, plants and animals are alike in patterns of physical growth, animals and man in sensual perception, man and angel in spirituality, and angel and deity in immortality. There exists in nature a universal bond through which all higher

powers—streaming downward in a long and continuous chain—communicate themselves to even the least significant things.

Such a universal order is perceived by Faust[13] in the sign of the macrocosm:

Wie alles sich zum Ganzen webt,
Eins in dem andern wirkt und lebt!
Wie Himmelskräfte auf-und niedersteigen
Und sich die goldnen Eimer reichen!
Mit segenduftenden Schwingen
Vom Himmel durch die Erde dringen,
Harmonisch all das All durchklingen!
Welch Schauspiel![14]

Goethe was familiar with many aspects of occult philosophy, and he expresses in beautiful verses what Agrippa describes in prosaic language.

The bond connecting heaven and earth and all things therein is the World-soul. Since this soul is a natural phenomenon, there also exists a "natural magic." The magnet, for example, is a magic object with occult properties by which it attracts iron. Since all things are connected either by sympathy or antipathy, Agrippa's work necessarily takes on encyclopedic scope. It gives a detailed account of all the magical effects of stones, herbs, animals, and stars as they have been described in books of this kind since the time of Pliny, conveying the idea that in a way everything is imbued with magic. Agrippa also writes on theurgical magic, which deals with demons and their conjuration.

Cardano knew the two works by Agrippa but thought little of them. On the basis of Agrippa's horoscope, Cardano characterizes him in the *Liber de exemplis centum geniturarum* as follows[15]:

Since the beginning of Virgo was his ascendant, with Mercury himself being the ruler of the house, of the exaltation and of the trigon, and with stars of mercurial and venerean nature also present, we can deduce that this man had an utterly mercurial character, that is to say, he was ingenious, erratic, untruthful, deceitful, but also studious. Thus he wrote a book on occult philosophy in which he compiled a thousand irrelevancies but offered no factual information. Such a book ought to be burnt because it leads people astray. The falling lunar node in the ascendant furthermore signifies a keen mind, clever but undisciplined, malicious and ill-intentioned. Hence he wrote about the futility of the sciences, and demonstrated—aside from the irrelevance of his subject matter—his own ignorance by discussing things he did not understand. Of course, many people enjoy this, just as donkeys enjoy their straw.

Cardano—in sharp contrast to Cornelius Agrippa, who was basically a charlatan—is a scholar to be taken seriously. Certainly, his mode of thinking cannot be called scientific when judged by standards of the seventeenth century or those of our own time. He was striving for a philosophy that would embrace the amazing multiplicity of phenomena and organize them into a meaningful whole. Such a philosophy derives its unity essentially from the stature and originality of its author. According to Jacob Burckhardt[16], the discovery of individuality was a major characteristic of the Renaissance. Insofar as he offers a concept of the macrocosm as a mirror image of the microcosm of his own self, Cardano is a typical representative of his age. This accounts for the many references to his own person and his experiences which we find throughout his writings. To Cardano, his scientific work is a means of understanding the world and human nature in general, as well as a way of gaining self-knowledge. He looks to science as an aid to orienting himself in a complex and dangerous world while being unable to effect any change. He wanted this analysis of the world and of himself to be scientific and rational; at the same time, he wanted it to satisfy the imagination as well as offer a way to comprehend the universe as a meaningful whole.

Modern scientific thought can no longer address itself to such a proposition, particularly since the idea that God's wisdom is revealed in the laws of nature—a belief which still inspired Galileo as well as Newton—has been discarded. In view of the incredibly unstable social conditions of the time, the goal Cardano set for himself was a highly difficult one, but his particular way of thinking enabled him to reach his objective with some measure of success. This we may at least infer when we see how Cardano, in the course of a long and hard life and despite great misfortunes, always regained his courage and philosophical calm. The great and lasting success of his writings shows, moreover, that his view of life also provided a real support for many of his contemporaries.

In attempting to describe Cardano's philosophy, we are opening to the reader that body of ideas which gave birth to our own world. The world of the sixteenth century is unfamiliar to us. But perhaps—by way of contrast—the difference between that world and the world of today will become apparent and will help us to understand ourselves. In addition, I wish to acquaint the reader with an important and extraordinary man who has continued to fascinate people.

Cardano's place in the history of science as a mathematician is assured. His book on algebra, the *Ars Magna*, was published in Nuremberg in 1545. Osiander, who in 1543 was instrumental in the publication of the work of Copernicus, also introduced Cardano to his publisher. An annotated English translation of the *Ars Magna* is available, to which I want to make express reference.[17] We shall focus our interest on Cardano the physician, natural philosopher, and psychologist.

1 Cardano's Life and Writings

OUR KNOWLEDGE of Cardano's life and work comes primarily from his own writings. They abound with accounts of his personal experiences and turns of fortune. These serve him as examples for his teachings or as a means of showing that his doctrines are based on his own experiences. Unlike those of many of his contemporaries, Cardano's personal statements have always proved to be truthful, insofar as they can be checked. His love of truth, in which he took pride, is beyond doubt.[1]

With the aid of this autobiographical material, Henry Morley[2] wrote a vivid and sympathetic account of the life of Cardano. At the age of seventy-five Cardano himself wrote *The Book of My Life (De Vita propria liber)*.[3] This is, however, not so much an autobiography as it is a self-portrait; it is a unique document of captivating liveliness. Jakob Burckhardt's[4] description of this work is unsurpassed:

> Whoever reads this book will be bound to its protagonist until he has finished it. Cardano does confess to having been a perfidious gambler, vengeful, hardened against all remorse, deliberately insulting in speech—but he acknowledges this with neither impudence nor humble contrition nor in an attempt to draw interest on himself. Instead, he is guided by the simple, objective sense of truth of the natural scientist. Most shocking

of all is, of course, that the seventy-six-year-old man who lived through most horrendous events[5] with a deeply undermined confidence in his fellow men, nonetheless considers himself reasonably fortunate: after all, he has a grandson living, he still possesses his vast knowledge, he is enjoying prosperity, high rank and esteem, he has powerful friends, mysterious things are revealed to him, and most fortunate of all—he has his belief in God. As an afterthought he counts his teeth: there are still fifteen.

A wisdom which, in its way, is quite grandiose—although not everyone will benefit from it—fills this book of life. Cardano acquired it at great pains in the course of a difficult life; he was a problematic character and lived at a time of many dangers and great cruelty. Even his contemporaries considered him an odd figure in many respects. But as Henry Morley correctly points out: "His eccentricity consisted perhaps more in the extent of his candour than in peculiarities of conduct and opinion."[6]

Subsequent ages were often unable to understand him. Gabriel Naudé (1600–1653), who published the *Vita propria* for the first time in 1643, draws in his preface a sketch of Cardano's character where his traits are unquestionably distorted toward the pathological. On the other hand, he holds Cardano the scholar in high esteem, noting: "Not only was he beyond dispute an outstanding physician, he was also probably the first and only man to distinguish himself in all sciences at once. He is one of nature's illustrations of what man is capable of achieving. Nothing of significance was unknown to him in philosophy, medicine, astronomy, mathematics, history, metaphysics or the social sciences, or in other more remote areas of knowledge. He, too, erred, of course, that is only human; it is marvelous, though, how seldom he erred."

Since Naudé's preface was placed at the beginning of Spon's edition of Cardano's works, his opinion was greatly responsible for the judgment future generations formed of Cardano's character, leading some people to believe that he was half crazy. But this only attests to the great change in the intellectual climate that occurred within only fifty years of his death, particularly in France. The fantasy world of the Renaissance was determinately left behind. Naudé, connoisseur of the literary world and "esprit erudit," friend of Gassendi, could not comprehend how so many opposites could be united in a single human being. He himself would probably have been unable to deal with such tension and would have gone insane—but then, he was no Cardano.

Girolamo Cardano was born in Pavia on the twenty-fourth of September 1501.[7] His father, Fazio Cardano, was an elderly and somewhat peculiar solicitor who had entered into a liaison with a much younger widow. It is uncertain whether this union was a clandestine marriage or not. If so, the father wanted to keep it secret out of fear of exposing himself to ridicule for getting married so late in life. An attempt was made to terminate his mother's pregnancy, and this may have adversely affected Girolamo. The parents lived separately in Milan, and the child was at first reared in the country. Later, the parents established a common household, brought Girolamo to live with them, and were considered married. Cardano says of his parents[8]:

My father dressed in a purple cloak in the old urban tradition, and he always wore a small black skull cap. He tended to stutter, he had whitish-grey eyes and could see very well at night, and up to the very end of his life he did not need glasses. My mother was short-tempered, had a good memory and mind, and was fat and devout. Both parents lacked consistency and constancy in their love for their children. It appears though—if one may say this—that my father was a better and more loving person than my mother.

Cardano loved and respected his father all his life, although he suffered some tyrannical treatment from him and was often thrashed when he was a young child. He was a feeble boy, fell seriously ill several times, and he was—either because of this or because of the aforementioned detrimental circumstances of his birth—more or less impotent until the age of thirty.

His father would have liked Girolamo to enter the legal profession. But Girolamo wanted to study medicine, a discipline which seemed to him—to use his own words—"purer and more sincere since it is based on reason and on the eternal laws of nature rather than on the transient opinions of men." His father was very distressed, "since he considered jurisprudence a nobler discipline aside from being much better suited to the acquisition of wealth and influence, and—most importantly—to the advancement of the entire family's position." But Girolamo persisted in his wish and pursued his medical studies at Pavia and Padua. While he was at the university, his father died. He took the degree of doctor of medicine at Padua in 1526. He had already met with difficulties here, probably in connection with doubts raised as to the legitimacy of his birth.

These were miserable times. Upper Italy was ravaged by war and pestilence, and Cardano was glad to settle down to the practice of medicine in the small town of Sacco near Milan. His physical condition improved to the point that he could get married. After several miscarriages his wife bore him two sons and a daughter. He never had any doubts that he would gain fame, if not wealth and power. He wished to perpetuate his name through his literary works.[9] It was difficult, however, to find a publisher. At Milan, he applied for membership in the College of Physicians, but without success: he was rejected because of his illegitimate birth. Cardano often lived in severe poverty, escaping from his anxieties by indulging his passion for gambling. His inquiring mind remained alert during such activities as well. His practical knowledge of gaming led him to discover one of the fundamental laws of the theory of probability.[10] It seems he was the first to realize that there is a "theory of chance." In 1534, through the intervention of influential friends, he obtained the position of physician at Milan's poorhouse (the Xenodochium), as well as a lectureship in mathematics at the Academy of Milan. Shortly thereafter he was offered a professorship in mathematics at Pavia, conditional, however, on a period of probation, and moreover offering very little pay. Cardano refused the offer as being below his dignity. His financial situation remained precarious. During these troubled times he wrote on astrology and on fate *(De Fato)*. He also wrote a book on consolation *(De Consolatione)* which was published in 1537, although it did not attract much attention at the time. In *De Libris propriis* Cardano adds this melancholic commentary to his report on that book: "All things human are transitory and no more than a moment's breath; and even our happiness is like effervescent water. There is no panacea: the troubles of the soul are altogether incurable."

In 1539, through the mediation of Osiander, who also assisted in the publication of Copernicus's book, Cardano found a publisher in Nuremberg who printed his *Arithmetica*. This book became a great success. Petreius, the publisher, wrote Cardano that he would print everything he could send him. The first thing he did was to reprint *De Consolatione,* which now found many readers.[11] Also in 1539, Cardano at last succeeded in being received into the College of Physicians at Milan. By a series of successful cures he attracted increasing attention and won influential friends. As a result, he was offered the chair in medicine at Pavia, which was a respectable offer, although not very promising financially. Cardano accepted the offer after lengthy consideration because his house collapsed on him. He had obviously still been living under rather desolate conditions.

During the subsequent years he lived sometimes in Pavia, sometimes in Milan, because the university often could not pay its faculty's stipends. It was at this time that Cardano wrote the *Ars Magna*, his book on algebra, which was printed in Nuremberg in 1545. This book made him truly famous, but it also entangled him in a vehement and unpleasant dispute with Niccolò Tartaglia. Cardano did not, however, carry on this dispute himself, but left it instead to his devoted pupil and collaborator, Ludovico Ferrari.

The cause of the dispute was this: Cardano had, among other things, solved the thirteen basic forms of the cubic equation,[12] and he had proved that his method would—at least formally—always provide a solution. He proceeded from a rule applicable to a special case which he had learned from Tartaglia: it is the equation $x^3 + ax = b$. Cardano discusses it in chapter eleven, introducing the subject as follows:

> Scipio Ferro of Bologna found the rule for solving this equation more than thirty years ago and communicated it to Antonio Maria Fior of Venice. Fior's competition with Niccolò Tartaglia of Brescia offered Niccolò the opportunity to rediscover it.

Upon my urgent plea he passed it on to me, although without proof. Armed with this support I worked out a formal proof ("demonstrationem quaesivimus, eamque in modos redacta sic subiecimus"), this was most difficult.

Tartaglia, on the other hand, maintained that in 1539 Cardano had solemnly sworn never to publish this rule, which Tartaglia regarded as his secret. He had, therefore, perjured himself. Ludovico Ferrari denied that his teacher ever took such an oath. It is difficult to discern where the truth lies. One thing that we know for certain is that in 1544 Cardano and Ferrari went to Florence together. On their way, they visited the scholar Annibale de Nave at Bologna, who showed them a manuscript of the deceased Scipio Ferro. In it they found Tartaglia's rule. Ferrari's account is credible since Ferro's manuscript is still at Bologna.[13]

In order to form any opinion on this matter one should take into account that Tartaglia[14] was a man of obscure origin; not even his family name is known. "Tartaglia" means "the stutterer" and is a nickname. When the French pillaged Brescia in 1512, his mother sought refuge for her son in the church. But the soldiers also invaded the sanctuary, and the twelve-year-old boy was severely wounded by a sword cut: his jawbone was split, causing permanent damage. With enormous energy—and for the most part autodidactically—he worked to become a respected mathematics teacher and mechanical craftsman. He lived by his skills and regarded his knowledge as his personal property. He was not a scholar in the true sense, but he had great practical knowledge—in applied mathematics as well as in mechanics—and he was a very successful teacher. Cardano, on the other hand, was a highly esteemed physician as well as a universal scholar, although he had not yet come to fame. The two men were of totally different mentalities.

When Cardano learned that Scipio Ferro had known this rule thirty years before Tartaglia's rediscovery, he probably no longer felt bound by his oath—if indeed he had ever taken such an oath—since it had been based upon false assumptions.

Cardano's encyclopedic work, *De Subtilitate,* was published in Nuremberg in 1550. He had been induced by dreams to write such a book; it became his greatest literary success. In 1554, an enlarged and improved edition could already be published—this time in Basel—and a third edition came out in 1560. In addition, the book went through about ten reprintings during Cardano's lifetime—in Basel, Lyons, and Paris.[15] As a result, *De Subtilitate* is generally regarded as Cardano's chief work. But as it is directed at a larger audience, it does not give us a complete picture of his own philosophy. In it, for example, he adheres much more closely to Aristotelian doctrine than he does in his actual philosophical treatises *De Uno* and *De Natura,* but then these were also written at a later date. Nonetheless, *De Subtilitate* bears in all respects the imprint of his personality and his comprehensive knowledge. In this he surpasses all his predecessors since Albertus Magnus as well as all his successors. In it, he also assumes a much more critical attitude with regard to all those wondrous occurrences usually recounted in such books.

In 1551, because of the threat of war, the entire university faculty left Pavia. Cardano went to Milan. There, a request reached the renowned physician to travel to Scotland to attend John Hamilton, Archbishop of St. Andrews. The archbishop, an important political figure, suffered from severe, regularly recurring attacks of asthma, often causing him to fear suffocation. Famous physicians had tried their skill, but so far to no avail.

Cardano reports on this case in chapter forty, entitled "Successful cures," of *De Vita propria:*

> The next successful cure was that of Archbishop Hamilton of Scotland who, then in his forty-second year, had been suffering from asthma for ten years. He had first consulted the physicians of the King of France, and then the physicians of Emperor Charles V, but to no effect. The patient was under the care of his personal physician—the Spaniard Casanatus—who was guided by the council of the Parisian physicians. Casanatus was reproached for the lack of improvement in the patient's condition, and I was finally obliged to express my opinion as to the reasons for his continued poor condition. Thereupon the archbishop became indignant with his physician and the latter with me because I had uncovered the trouble. So I was apprehensive of the one, and accused by the other of procrastinating the cure—especially since he improved as soon as I began to direct the treatment. In the midst of this uncomfortable situation I asked permission to withdraw from the case. The archbishop did not react graciously to my request, but he did allow me to leave. Upon leaving, I prescribed a detailed regimen, with which the patient regained his health within two years. I stayed with him for seventy-five days. His cure has been attested.

Cardano obviously faced a difficult and delicate task. Not only had the patient been seriously ill for years; he was also surrounded by physicians who, in Cardano's opinion, were prescribing the wrong treatment. In order for him to be more successful, Cardano had to convince both the patient and the attending physicians to trust his better judgment without unduly offending his colleagues. He proceeded, therefore, with utmost discretion, confining himself to carefully observing the patient's way of life and physical condition. Many weeks passed until Cardano finally decided to assume an active role. His intervention was evidently crowned by success.

Upon his departure, Cardano left behind a detailed scientific analysis of the nature of the illness and its proper treatment. He also gave the archbishop precise instructions in writing for a mode of life that would ensure his recovery. Both documents are printed among the *Consilia Medica.*[16] I shall discuss these consilia in the following chapter to give the reader some idea of the way in which Cardano practiced medicine. Suffice it for the time being to note that Cardano's directives were most reasonable. But most importantly, he made such a strong personal impression on the patient that the latter strictly adhered to the medical instructions. Two years later, in October of 1554, the archbishop expressed his gratitude in a letter to Cardano.[17] The archbishop reported that he had recovered his physical strength and had gained a new lease on life. The attacks were now two months apart and much less severe. He attributed this to Cardano's excellent instructions and to the great pains he took over him.

The archbishop's illness seems to have been a case of a complicated psychosomatic allergy. One must remember that in those days the life of an important public figure was strenuous as well as dangerous. This will be apparent to anyone familiar with the political situation in Scotland at that time, which greatly affected the position of Archbishop Hamilton.

In 1542, James V, King of Scotland, had died at the age of thirty after an unhappy reign. He left a daughter, Mary Stuart, who was ten years old in 1552. She was engaged to the French crown prince Francis II, and was educated in France—she never really learned to speak English properly.

James Hamilton, Earl of Arran, was Regent of Scotland, frequently being engaged in multifarious disputes with the widowed queen, Mary of Guise. The archbishop, an illegitimate child, was the earl's half-brother. He had become keeper of the privy seal in 1542 and archbishop in 1546. This meant deep involvement in the regentship, and since its inception he had been suffering from his severe attacks. Cardano did cure him, but twenty years later the archbishop died under tragic circumstances. He was accused of complicity in the murder of Lord Darnley, husband of Mary Stuart. He was found guilty and hanged.

For Cardano, the journey to Scotland was an exhilarating experience, but he never wrote down a complete account of it. However, in *De Rerum Varietate* he reports on many of the things he observed and found noteworthy. I have put these scattered notes into chronological order and thus composed a kind of travel book.

After Cardano had agreed to offer his medical services to the archbishop, he received the sum of two hundred crowns[18] to cover traveling expenses to Lyons, where Casanatus, the archbishop's personal physician, was going to meet him.

Cardano left Milan on the twenty-third of February 1552. He rode via the Simplon pass, the Valais and Geneva to Lyons where he arrived three weeks later, on the thirteenth of March. Casanatus was not there to meet him, but this did not cause Cardano any concern. Scotland was far away, the crossing of the Channel in winter was often impossible for long periods, and traveling was generally a troublesome and precarious undertaking, although this did not discourage people from traveling extensively.

Cardano could see evidence of the ties between Scotland and France already in Lyons. He noticed that bituminous coal from England and Scotland was used here in the processing of iron. This was interesting to him, because in Italy only charcoal was known at that time. This explains why the coal he saw in Lyons did not seem like real coal to him. He says (*RV,* 88)[19]:

> I have seen stones that can be ignited and that burn. They are called "coal." I saw them in Lyons. They look like ordinary coal, that is to say charcoal, but they are heavier and they give off a bad smell. Only blacksmiths use them in the forge because they produce a very intense fire. These stones are imported from Scotland and England. Those which are pure are quite light and shiny. I saw a relatively small cart filled with this coal being drawn by eight yoke of oxen. This is, however, not an indication of the weight of the stones, but rather of the weakness of the oxen. The fact that they are imported from so far away is evidence that they have been found exceptionally useful in the processing of iron.

As regards Cardano's comment on the large yoke of oxen, I feel that he misinterprets the reason for it. I would think that it is rather an indication of the incredibly poor condition of the roads which, particularly during wet weather, turned into veritable loam-pits where the wagon wheels got stuck.

At last, Casanatus arrived in Lyons, bringing with him an additional three hundred crowns, and on the first of May the party set out for Paris. As was customary, most of the journey was completed on horseback. But when Cardano reached the Loire, he hired a boat for himself and his horses to travel downstream to Orleans.

While in Lyons, it had become apparent to him that this would be a journey of considerable duration, and that he ought to find something to do in order to pass the time. In Lyons he had come across Ptolemy's famous textbook on astrology, the so-called *Tetrabiblos,* and he decided to take advantage of the time the journey afforded him to write a commentary on this classical work. Like many physicians of his time, Cardano was himself an eminent astrologer. And just as he strived in his medical practice to move away from medieval-arabic medicine and return to the original authorities, that is Hippocrates and Galen, he found Ptolemy's reflections on astrology far more relevant than all the speculations of his Arabic successors. One can picture him on the ship among his horses, reading Ptolemy and jotting down his comments. ("Epistola Nuncupatoria" to "Cardani in Cl. Ptolemaei de Astrorum Iudiciis Commentaria," dated July 1553.)

There was a further prolonged sojourn in Paris. He met with other physicians who were attending the archbishop and who were anxious to hear his opinion on this extended, severe, and mysterious illness. It was quite common then for a medical consultant to make his diagnosis purely on the basis of a description of the symptoms, and then recommend a treatment. Cardano himself did this frequently, as evidenced by the numerous medical consilia included in his published works. However, in the case of the archbishop, this practice seemed too unreliable. Cardano expressed his wish to see the patient himself, and did not reveal his opinion.

In Paris he made the acquaintance of many an eminent contemporary, among them Ranconet, president of the "chambre des enquêtes"[20] whose horoscope Cardano cast.[21] Ranconet was a highly educated man, and Cardano also liked him personally. Cardano warned him of impending danger; he seemed to be facing a violent death. Ranconet indeed died under tragic circumstances only a short time later. From Paris Cardano went to visit the church of St. Denis where the French kings are buried. He writes (*RV*, 672):

While in Paris, I was fortunate to meet John Manien, an impressive man and important mathematician [like Cardano himself]. He came to see me daily and was in every way most obliging. He happened to be physician to the monks of St. Denis, the world famous church three miles outside of Paris, and he took me to see this most holy place. After I had visited the royal tombs and had seen the statues and marble ornaments, the horn of the unicorn, which hangs in the church, was lowered for me and I could get a very close look at it. It is so long that I could hardly reach its tip when it was put upright. Considering its length it is surprisingly thin. I could easily clasp it with my thumbs and index-fingers. It tapers almost imperceptibly toward the tip where it is still as thick as a thumb. It is perfectly smooth and marked as if by a fasces which moves upward in a spiral pattern. There are five rods, separated by fine lines, which also spiral upward. It is hollow like the horns of cattle, but the hollow part is no bigger than the solid part. It is perfectly straight. The color is that of antlers. It weighs seventeen and a third pounds.

Cardano also visited the church's treasure room (*RV*,677), where he particularly admired some bowls of agate, "which were distinguished even more by craftsmanship than by material and size." Cardano was an expert on and an amateur collector of carved precious stones. He also found especially noteworthy a sword set with jewels which had belonged to Bishop Turpin, who lived at the time of Charlemagne. "It was very light as is suitable for an old man." Another sword was the exact opposite; it was so heavy that Cardano could hardly remove it from its sheath. "It was rather short—the sword carried by Joan of Arc who once liberated the French from the yoke of the British. The weight of the sword is an indication of the maiden's unusual strength." Cardano also saw the coronation vestments. Finally, he was shown an object that looked like a cow's horn, but which was said to be the talon of a griffin. The foot was supposedly kept in the town hall nearby. Cardano is doubtful that any animal could have such large claws. It probably was a forgery, a cow's horn that had been hollowed out.

From Paris, Cardano continued his journey down the Seine to Rouen, at that time one of the major cities of France. To Cardano it was a paradigm of a beautiful town (*RV,* 667):

> It is well situated, it has fine architecture, and its inhabitants are handsome. It lies among hills and fertile fields, along a navigable river with beautiful islands. Unfortunately the region is a little too far north for a climate mild enough to grow grapes. The town itself has a stone bridge and stately houses and impressive churches. The spires of the cathedral are of extraordinary height and magnificence—I don't know if there are any more beautiful in Europe. One of the bells is of great size as well as mighty sound. It would seem that only Rome is even better situated and still more imposing and beautiful. Paris, on the other hand, although the largest city in France, is a rather filthy place, filled with stench and poisoned air. I wonder if the name "Lutetia" is a derivative of "lutum" meaning dirt.

From Rouen one travels up the coast by way of Dieppe to Calais, then still in English hands. The party made a stopover in Dieppe, and Cardano had the opportunity to see some interesting fish in the harbor of the fishery—among others, the porpoise and the thornback with its poisonous spikes (*RV,* 226). But what delighted him most on his walk through the town was a gooseberry bush (*RV,* 108). He knew the plant as a low-growing bush with small leaves the shape of grapeleaves:

> It is prickly and bears small round berries, which when ripe are greenish-yellow and have a sweet-astringent taste. This bush had, however, grown into a little tree and the berries were the

size of cherries. It seems that the sap in a bigger stem is of better quality and therefore produces larger fruit. And this in a climate too cold for vines to grow! Because the plant had branched out so much and had copious foliage it had been enclosed with wirenetting. Little birds were living in it. What a pretty sight! It will be difficult to find anything like it elsewhere.

No further mention is made of the remainder of the journey to Edinburgh. Cardano arrived there on the twenty-ninth of June, two months after leaving Lyons. He stayed until the thirteenth of September.

Scotland in those days was a poor and wild country. Cardano gathered detailed information on its geography and history. He bought the *Historia Scotorum,* by Hector Boece, probably in Paris, where it had first been published in 1530. Boece was a friend and fellow student of Erasmus, and he later became the first "regent" of the University of Aberdeen. His *Historia* was widely read, was translated into Scottish and English, and was also included in Holinshed's *Chronicles*. Here, Shakespeare read the macabre tale of Macbeth. Cardano read it in Boece, and relates it in *De Rerum Varietate,* page 537, together with similar stories. Scotland was then indeed a land of widespread cruelty and superstition.

However, he also learned from the archbishop of the many islands in the north of the country: the Orkney and Shetland Islands. The main source of nutrition in that region is air-dried smoked fish. The people do not drink, quarrel, or worry about tomorrow; they are tall, handsome, and healthy. This must be the land of the Hyperboreans, a legendary ancient people (*RV,* 18).

The severe climate has a marked effect on animals as well as people. On those islands there are small horses hardly bigger than donkeys. The cold also stimulates hair growth: "I remember seeing a younger man in England whose front-body was to my amazement completely covered with hair. In Scotland I saw another man who had hair over his entire body; one would have called him a big bear."

To Cardano the country appeared extraordinarily cold and windy, and it was hard for him to imagine living in a country without vineyards. This is reflected in the following observation (*RV,* 124):

It is marvelous that the plane-trees which are being neglected in Italy and are therefore seldom seen, are cultivated in Scotland and are quite common there. In the arboretum of the Augustine monks in the suburbs of Edinburgh I counted more than twenty of these trees, some of them over thirty feet high. I believe that the people there take pleasure in them because their leaves remind them of grape-leaves. They are so much alike that I thought the Scots were growing vines when I saw plane-tree saplings. In view of the fact that their land produces relatively few trees, the people were right to select the most beautiful for cultivation.

This observation is actually inaccurate, as Cardano is confusing the sycamore (acer platanoides) with the plane-tree, which—as the Latin name shows—can easily happen. And the leaves of the sycamore remind him, not the Scots, of the grapevines he has been missing so much. He also finds it most regrettable that olive and walnut trees do not seem to flourish either in Scotland or in England. He writes: "Although there is an ample supply of fat, especially butter, the combination of butter and fish causes elephantiasis, a disease which is widespread in France and England and particularly in Ireland."

This is an amazing medical observation. Cardano could indeed be referring to "elephantiasis arabum," which is caused by filial worms, since it is known that Ireland in particular was also infested with malaria at that time. Cardano must have been glad to leave the inhospitable country. In addition to the five hundred crowns for traveling expenses, the archbishop paid him a fee of eighteen hundred crowns, "of which 1400 actually came to me" (*Vita propria,* chapter 40). This was a very liberal reward. Cardano was also presented with a very beautiful horse, an ambler with white markings on which he rode home. He describes the animal in *De Rerum Varietate* and says that he is still riding it (*RV,* 188). The same work also contains a description of how a horse is trained for ambling (*RV,* 181). Horses thus broken in were considered the most comfortable way for elderly gentlemen to travel. (The horse of Donatello's Gatta Melata in Padua is also an ambler.) Cardano rode his ambler southward, first crossing the coal-mining region in the north of England. He reports (*RV,* 30):

> England is such a bituminous place that not only can stones be ignited there, but the soil burns as well. After the heather (erica) has been cut, the soil is stored in great heaps near Newcastle. It is used for fuel, and also in place of tile for roofing. The roots of the heather are used to ignite it. As we crossed the region on our way from Scotland to London, the ground was shaking markedly. I realized though that this was not because there were caverns under us but rather due to the fact that this black earth is porous, like a dry sponge. Combustible earth and stones are so abundant here that they are much less expensive than our charcoal at home. However, the soil is watery and therefore produces much more smoke than the stones. Frequently, these are so dry that they hardly smoke at all.

In London, Cardano had made the acquaintance of Sir John Cheke, the tutor and advisor of the young king. The two men became intimate friends. Through Sir John,

Cardano was granted an interview with the king, Edward VI, a youth of fourteen. Cardano thought him most congenial. The king's horoscope was printed in Cardano's commentary on Ptolemy as the first in a series of twelve illustrative genitures. Here Cardano remarks on the young monarch:

> He was a wonderful boy who, I was told, had already learned seven languages. He was as fluent in French and Latin as in his native tongue. He was trained in logic and was extremely intelligent. He was in his fifteenth year when I met him. He asked me, speaking Latin as beautifully and fluently as myself, «What new ideas does your book *De Rerum Varietate* contain?» (That was the title of my manuscript.) I replied: «In the first book I explain the most important causes of comets.»

Cardano proceeds to relate the conversation he had with the young boy and then continues:

> He was very open and most amiable, and he raised everyone's fondest hopes. It is most lamentable that a life holding such promise was snuffed out so soon, a grievous loss to England and indeed to the world. [Edward died in July of the following year, 1553]. He was so cheerful; he brought youth back to his teachers; he played the lute; he was interested in public affairs; and he was a free spirit, like his father.

The young king's father was Henry VIII, who among his contemporaries was not at all ill-reputed. On the contrary, this violent and egotistical man enjoyed a curious popularity, the reason being that he protected the common people, while those in high office had good reason to dread his terrible temper: he was a true king. Cardano is of the same opinion, and he believes the execution of Sir Thomas More to have been the king's only serious mistake. But this did, of course, have disastrous consequences.

When Cardano cast the boy-king's horoscope, he did not foresee his untimely death. On this he comments as follows:

> This was an act of fate—ecce fatum!—: it had taken one hundred hours to make this forecast for the king's future. If I had worked just another half hour, I might have had a presentiment of the danger: that is to say that his life was in danger, for I am not one to predict certain death. But then I would have been obliged to pronounce a judgment—and imagine what alarm that would have caused.

Cardano points out the dangers he would consequently have had to face in a foreign land and relates how badly other prophets of misfortune had fared. He continues:

> I could simply maintain, as other astrologers have, that I knew quite well what would happen, but that I kept silent out of fear. But that was not the case at all. I did not think of that, and I did not foresee anything. One thing, however, I could see quite clearly, although independent of any astrological indications: Everything was in the hands of one man—that is John Dudley, Duke of Northumberland—, the boy, the administration, the treasury, parliament, the fleet. His father had been executed by the king's father (that is, Henry VIII); he himself had two uncles on his mother's side condemned and executed, and as he purloined everything, in the end it was fear rather than hatred that induced him to conspire against the king's life.

When Cardano became aware of the situation, he was "seized by a feeling of dizziness; I was disheartened at the state of the kingdom, and being the better prophet on account of an innate good sense rather than astrological insight, I hastened to leave, filled with fear and forebodings of disaster." [Edward VI was the son of Henry VIII and Jane Seymour who died in childbirth. His uncles were Edward Seymour, Duke of Somerset and Lord Protector, who became regent after Henry's death, and his brother Thomas, Baron Seymour of Sudely, Lord High Admiral. The admiral plotted against the Lord Protector and was executed in 1549. This greatly hurt the Lord Protector's popularity. He was subsequently deposed by John Dudley, a statesman who had reorganized the system of taxation and revenues from feudal lords for Henry VII. This had made him quite unpopular. After Henry VIII ascended the throne, Dudley was charged with treason, and in 1510 the king offered him up to the scorn of the people. A son of the Duke of Northumberland is Robert Dudley, favorite of Elizabeth.]

Cardano had detailed knowledge of all these violent events, although he falsely ascribed the death of the admiral to the Duke of Northumberland, just as Shakespeare attributed the death of Clarence not to Edward IV, but to Richard III.

With his Scottish horse and accompanied by a young secretary he had engaged in England, Cardano traveled in October to the Netherlands, where he visited the cities Bruges, Ghent, Brussels, Louvain, Mecheln, and Antwerp.

In Brussels (*RV,* 429) he heard chimes for the first time, a carillon played on the churchtower. He found the affair admirable rather than pleasing, since the music was not at all melodious: "I saw these carillons also in Louvain and in Antwerp. Their purpose is to give everyone in the city a chance to hear the music, and that is in itself something well worthwhile."

Also in Belgium he apparently saw red cabbage for the first time (*RV,* 127): "The heads are purplish in colour, almost a violet-blue, a cheerful sight! It is amazing that a common vegetable could be so attractive. However, white cabbage is said to be the better vegetable."

From Belgium, Cardano proceeded by way of Aix-La-Chapelle, Cologne, Mainz, Worms, Speyer and Strasbourg—where the cabbage was particularly large—to Basel. There he stopped again for a longer sojourn, as he had previously done in Antwerp.

Aside from being a university town, Basel was above all one of the most important printing centers in the world. Cardano's publisher since 1543 had been Joh. Petreius in Nuremberg. But he had since died, as noted in the preface to the Basel edition[22] of *De Subtilitate* (1554). Besides, Nuremberg was far from Milan; but most importantly, the political situation in the empire was most unstable. An obvious alternative for Cardano was, therefore, to look for a publisher in Basel. He found him in Henry Petri who, in 1541, had published the famous cosmography by Sebastian Munster, a work which went through innumerable editions into the seventeenth century. Petri seems to have been the ideal publisher for Cardano. He must have been very happy about the prospect of an association with the renowned physician and scholar, especially since Sebastian Munster, his most eminent "success author" until that time had died in May of 1552. Petri published Cardano's commentary on Ptolemy in 1554. From then until 1570 Cardano published a new work with Petri almost annually. In 1557, *De Rerum Varietate* was printed in two editions: one in folio, and one in octavo, the latter as a "low-priced" edition.

Cardano was an ardent admirer of Erasmus. Although Erasmus had died in Basel fifteen years earlier, Cardano must have met people there who still remembered him well. The commentary on Ptolemy, which was composed during this journey, presents as the twelfth example in the selection of genitures the horoscope of Erasmus. Cardano says of him: "Fuit exigui corporis et macilentus, iucundus ob venerem in ascendentem ac lepidus." (He was of small, delicate build, an amiable person, as signified by Venus being in the ascendant—a man of most graceful and charming wit.)

There was one particularly famous sight in Basel: the oak of the Petersplatz (*RV,* 123), which Cardano calls "divi Petri Nemus."[23] This is how he describes the tree:

> Three men with outstretched arms can hardly clasp its trunk. The trunk is less than man-high. Six or seven boughs slant upward from its upper part, and form a kind of chalice. Many small branches and twigs make it very thick. I was told that Emperor Maximilian once took a meal and an after dinner nap in the tree's crown. This is indeed feasible as the circumference of the crown is a good one hundred feet; I found this out myself by walking around it. Due to meticulous care, the Basel oak tree rivals the Lycian plane tree. Lucius Mucianus boasted that he frequently feasted with as many as eighteen companions underneath this plane tree, but the German oak can host more guests in its crown than the plane tree in its shade. The extraordinary thing though is this: The plane tree Mucianus wrote about was eighty-one feet high, whereas the oak tree is exceptionally low.

Cardano regarded this oak tree as a paradigm of arboriculture, as a horticultural work of art, since no oak tree would naturally grow in this form. In the true spirit of his time, Cardano likened it to the tree mentioned by Pliny in the twelfth volume of his encyclopedia, and comments: "Whereas the plane tree of Mucianus grew naturally, the artistry of the Basel horticulturist produced something much more remarkable."

It must have been late in the year when Cardano left Basel. He proceeded to Berne via Neuenstadt, probably intending to return from there via the Vaudois and the Simplon pass to Italy. In Berne he apparently received an unexpected invitation to Besançon and accepted it. It seems that Emperor Charles V wished to engage him as personal physician and requested him to enter into negotiations with one of his representatives. Besançon was a suitable place for this, as it was a free city as well as being favored by the emperor. He granted the city to bear in its coat of arms the imperial device: the columns of Hercules with the words *plus ultra.* A century later, the city still minted coins with a full length portrait of the emperor. Cardano received several offers during this journey, as he reports in chapter thirty-two of *De Vita propria.* He apparently declined all of them without much hesitation. Of course, in the case of an offer from the emperor it was not that simple, especially since Milan was under Spanish rule.

It seems that the Bishop of Lisieux[24] received Cardano in Besançon as representative of the emperor. This is curious, because Lisieux is in Normandy. But then one has to remember that at that time the Church was internationally administered and the bishops frequently resided outside their own diocese.

Cardano stayed in Besançon for quite some time; the bishop was hospitable and presented his guest with many gifts. But Cardano could not resolve to enter the emperor's service. The emperor would certainly have benefited from the counsel of a personal physician; for the past twenty years he had been suffering from severe attacks of arthritis: his unreasonable dietary habits were almost proverbial[25] and did not help matters. At the age of fifty-two, he was practically an invalid. On October nineteenth he had laid siege to the city of Metz which the French, supported by the Germans, had seized. Although the emperor and the famous Alba were personally in command, the siege had taken a turn for the worse. Under these circumstances

Cardano felt that the prospective position was just too precarious, "since the emperor had gotten into a most difficult situation, having lost the major part of his army through cold and starvation."

Cardano decided to return home. He could no longer take the route over the Simplon pass since it was now December and the last escorted party of travelers crossing the pass had left. One must realize that no one traveled across the Alpine passes on his own; rather, one joined a group which was organized and led by local guides. In the seventeenth century, these travel groups were organized from Brig, by the family of Stockalper. They amassed a fortune since they controlled the travel route from Upper Italy all the way to Lyons.

Cardano decided to travel via Zurich and the Splügen, which was still possible (*RV,* 508). Murer's 1576 map of Zurich shows in detail what the city was like when Cardano passed through it. He paid a visit to Conrad Gessner, town physician and professor at the Carolinum. They discussed distillation methods and prescriptions, some of which Cardano published in *De Rerum Varietate.* In Italy Gessner was still hardly known, because only the first volume of his famous animal books published between 1551 and 1587 had so far been printed. Cardano recognized Gessner as an important man and colleague and wanted to further his reputation. With this purpose in mind, Cardano wrote (*RV,* 508):

> No doubt people will wonder why I am quoting Gessner and they may say that "mules mutually rub". I shall explain my reasons. On my way through Zurich I made the acquaintance of this man. He is humane and modest, in keeping with his dedication to the muses. He is highly learned, and his interests extend beyond the mere body of knowledge (sapientia!) to anyone striving for knowledge. He wrote down the above

recipe without adding his name and gave it to me as a present. I am not asking why he omitted his name; I just want to express my admiration for the generosity he accorded to a stranger.

Cardano furthermore praised Gessner as a writer, judging him to be superior to most, including himself. But above all, Cardano wanted to draw attention to Gessner's sincerity and honesty, which are worth more than all erudition (*RV,* 214).

Aside from the congenial Gessner, Cardano found another curiosity worthy of note, the remains of a pelican nailed to the door of the Zurich town hall. He gives a detailed description of the bird—he is bigger than a swan, with a very long beak and black flat swan's feet:

> The bird was killed in the Lake of Zurich which is near the town and from which the river Limmat flows. The animal in question is most likely a waterbird that devours great quantities of large fish, forces them back, eats what he likes, and then leaves the bones. As the name Onocrotalus indicates, the bird's voice—or rather the great noise he makes—is similar to that of a donkey. The bird is thought to be an oceanbird that occasionally migrates to the lake, undoubtedly for good reasons.

From Zurich, Cardano went on to Chur, from where we have a last travel report. Where Cardano talks about cities and buildings he has seen, he says (*RV,* 699): "The most beautiful private house I saw was that of the steward Jacobus, just a stone's throw from Chur. One of the most splendid sights in this house was a beamed wooden ceiling painted in the most exquisite natural colors. One cannot describe it if one is not actually looking at it and only the most eloquent and faithful description and illustration could do it justice."

I am somewhat surprised that it is a house outside of Chur that Cardano singles out as being especially beautiful. Perhaps it just seemed so in retrospect, as Cardano was at that point still facing the hazardous journey from Chur to Chiavenna, in the middle of December. But it must have gone well, since at the beginning of the New Year Cardano was back in Milan.

Cardano's medical practice was highly successful, and his economic situation had improved considerably. His oldest son, Gianbattista, took his doctoral degree in medicine and, to the great satisfaction of his father, was admitted to the College of Physicians. But subsequently (and despite Cardano's most vehement opposition) he married a woman of ill repute. Several instances of infidelity drove Gianbattista to poison her. This had catastrophic consequences. The father himself conducted the defense of his son at the trial, but the vengefulness of the victim's family triumphed, and on the seventh of April 1560 the son was executed in prison. Cardano barely recovered from this blow. Added to his deep grief was the social defamation that affected the whole family. Only with great mental effort was he able to gather enough strength to see even such misfortune as ordained by God.

Still, he had retained some powerful friends, among them the Cardinal Archbishop of Milan, Carlo Borromeo, one of the most distinguished figures of the counterreformation. Cardano assured himself of the cardinal's lasting gratitude by curing his mother of a serious illness. It was primarily due to the cardinal's mediation that in 1562 Cardano was appointed professor of medicine at Bologna, then the second capital of the Pontifical state. Here, Cardano was held in high esteem and named an honorary citizen of the town. He had influential friends and enjoyed great success as a teacher. Of course, he had to endure the calumny of jealous colleagues, but he was able to put himself above that and was not harmed by it.

In 1570, however, new storm clouds gathered above his head. He was arrested by the Inquisition, apparently being accused of heresy. No details of the charges are known; Cardano made no reference to the matter. Yet, there are several reasons why such a charge could have been brought, as the attitude concerning matters of faith had recently grown rather uncompromising. It was due to the diplomatic skill of Cardinal Morone that in 1563 the council of Trient came to an end with the curial interests prevailing. In 1566, the affable and high-spirited Paul IV was succeeded by the somber figure of the ascetically devout Pius V, who was determined to enforce the reforms of church and dogma. Ranke[26], the historian, says of him:

> He was not satisfied that the Inquisition requited current transgressions; he had cases reopened that went back ten, even twenty years.
>
> He insisted most rigorously on the way the ordinances of the church should be administered. In one of the papal edicts we read: "A physician who is called to a bedridden patient is forbidden to attend to him for more than three days unless he receives at this time an attestation that the sick person has confessed his sins anew."

If the Inquisition officials put their minds to it, they could certainly discover clandestine heretical statements in the works Cardano had published over the past twenty years. In *De Subtilitate* he put Christianity, Judaism, and Islam side by side and objectively compared them. This led later commentators to believe that he was the author of the famous apocryphal book, *The Three Impostors.*[27] About 1540, Cardano had written a book *De Immortalitate animorum,* in which he compiled everything that had up to that time been said about the immortality or mortality of the soul—a question much discussed since Pompanazzi had published a book under that title in 1516.[28] Cardano's own doctrine concerning the soul could hardly be called orthodox either. He had, after all, cast Jesus's horoscope and published it in his commentary on Ptolemy.[29] It was suppressed in subsequent editions, and was only later inserted into Spon's edition of Cardano's works.[30]

The horoscope was undoubtedly a controversial issue, but at this point the cardinals Borromeo, Cesi, and Morone, president of the council of Trient, intervened on behalf of their friend Cardano. After nearly three months in prison—where he was incidentally quite well treated—he was released on bail. For the following three months he was kept under house arrest. Meanwhile, the inquest was taken over by authorities in Rome, and Cardano was obliged to move there as well. He arrived on October sixth, 1571, the day the victory at Lepanto was being celebrated. At first, he was filled with anxiety as to what was awaiting him there. But no further measures were taken against him, except that he was forbidden to publish. By order of the Pope he was received into the Roman College of Physicians, and was moreover given a pension. Cardano was now over seventy years old. He lived for another five years, practicing medicine, reading and writing.

In Rome he wrote his autobiography, in which he takes a close look at himself and the world in which he lived. There is a Chapter (31) entitled "Happiness" (Felicitas), where he says:

> Although happiness suggests a state quite contrary to my nature, I can truthfully say that I was privileged from time to time to attain and share a certain measure of felicity.—If there is anything good at all in life with which we can adorn this comedy's stage, I have not been cheated of such gifts: Rest, peace, quiet comfort, self-restraint, orderliness, change, fun, entertainment, pleasant company, coziness, sleep, food and drink, riding, rowing, walking, obtaining the latest news, meditation, contemplation, good education, piety, marriage, merry feasts, a good and well-ordered memory, cleanliness, water, fire, listening to music, beholding the beauty surrounding us, pleasant conversation, tales and stories, liberty, continence, little birds, puppies, cats, the consolation of death and the thought of the eternal flux of time as it flows past all—the blessed and the afflicted, happiness and misfortune; and then there is always the hope for some unexpected good turn of fortune; there is the practice of an art one is skilled in; there are the manifold changes of life, the whole wide world!

These reflections of the wise old physician shall conclude my account of his life.

As regards Cardano's literary output, his book *De Libris propriis*[31] offers an initial overview. Cardano's models for this work are Galen and Erasmus. He wrote it, after various preparatory writings, in 1562, and dedicated it to his friend Nicolaus Sicco, president of the court of justice at Milan. In it, he relates why and when he wrote his more important works, thus giving us an autobiographical sketch which supplements the *Vita propria.* In *De Libris propriis,* Cardano makes a special point to tabulate the sequence in which his books ought to be read. The table lists only a selection from his writings, but there are still forty-six works specified. The directive to read his works in a certain order indicates that Cardano thought of his writings as a whole, like a microcosm of his thoughts and his knowledge.

Not all of the works tabulated were published during Cardano's lifetime, so that this is really for the benefit of posterity. He also destroyed many manuscripts in his old age, as he reports in *De Vita propria,* because he was no longer satisfied with them. He did, however, keep excerpts of what he thought was worthwhile saving.

Karl Spon had access to the literary remains, although in some cases it was impossible to decipher the manuscripts. Some of them were spoiled, even eaten by worms. Yet, almost all the works listed in this table are included in the *Opera,* albeit sometimes in fragmentary form. The titles are often abbreviated in the table, but it is always clear which work is meant. I am reproducing this table, adding a number to each text to indicate the chronology of the works according to *De Libris propriis.* This order applies to first copies only, which were often revised, and notably enlarged, later on.

It is questionable if the chronology of the writings is a reliable indication of the development of Cardano's thinking. To take an example: *De Natura* was written ten years after *De Subtilitate* and is decidedly less Aristotelian than the earlier work. But this does not necessarily mean that with advancing age Cardano diverged more and more from Aristotelian doctrine. Perhaps he simply decided that it would be better to present a more conventional philosophy of nature in a work that was directed at a wide and diverse audience. Even within a single work his presentation is not consistent—his mind just did not work that way. He was well aware that there are many sides to every problem, and that there are as many solutions as there are points of view.

Still, a chronology of the genesis of Cardano's writings gives us some idea of the direction in which his philosophy developed.

Cardano's "Tabula" of Works	*Chronology of Composition*	*Volume, Opera*
1. Elementa Geometrica	1	IV
2. Arithmetica	5	IV
3. Musica	21	X (fragment)
4. De Uno	37	I
5. Dialectica	30	I
6. De Secretis, lib. primus	38	II
7. De Libris propriis (original version)	17	I
8. De Natura	46	III
9. De Subtilitate	22	III
10. De Rerum Varietate	27	III
11. De Gemmis et Coloribus	39	II
12. De Animi immortalitate	15	II
13. De Fato	3	
14. De Arcanis aeternitatis	13	X
15. Supernaturalium liber	47	
16. Theognostos	29	II
17. Hymnus	34	I
18. De Moribus	11	X
19. De Sapientia	14	I
20. De Praeceptis	24	I
21. De Utilitate ex adversis capienda	33	II
22. De Minimis et Propinquis	41	I
23. De Summo Bono	42	I
24. De Consolatione	12	I
25. Gulielmus seu de morte	44	I
26. Tetim, seu de humanis consiliis	40	I

Cardano's "Tabula" of Works	*Chronology of Composition*	*Volume, Opera*
27. Commentaria in Mundinum	36	X
28. Commentaria in Hippocratem	18	VII, VIII, IX
29. De Sanitate tuenda	23	VI
30. Floridorum	8	IX
31. De Urinis	32	VII
32. Ars medicanda parva	31	VII
33. De Experimentis	20	
34. De Aqua et Aethere	28	II
35. De Indico morbo	9	
36. De Morbis compositis	2	
37. Contradicentium Medicorum liber	16	VI
38. De malo Medendi usu	7	VII
39. Quod nullum Simplex	6	VII
40. Commentaria in Ptolemaeum	26	V
41. De Iudiciis astrorum	4	V
42. Metoposcopia	19	*
43. De Somniis	10	V
44. De Secretis lib. secundus	35	
45. Problemata	25	II
46. De Nodis	43	
47. Paralipomena	45	X

* Printed in Paris, 1658; not incl. in the *Opera*.

As is evident from the list, books one to six are propaedeutic texts, corresponding to the *artes liberales* of the traditional universities. Books eight to eleven contain writings on natural philosophy, followed by reflections on things supernatural and divine in books thirteen to seventeen. Books eighteen to twenty-six include the cosmology, writings on moral philosophy and the books of consolation. The medical writings are next, numbers twenty-seven to thirty-nine, filling four folio volumes in the *Opera*. Listed last are the texts on astrology, physiognomy and the interpretation of dreams—that is, writings pertaining to the mantic arts, which in Cardano's opinion only met the criteria of probability.

The first seventeen books are theoretical works; the subsequent ones are practically oriented, dealing with the proper way to live, therapeutics, the "diagnostic" arts. The *Hymnus* separates the theoretical works from those concerned with practical matters in a manner analogous to the distinction made in *De Libris propriis* between propaedeutics and the "higher" sciences—that is, "higher" in the sense of the upper academic faculties as opposed to the lower artistic faculties. The *Hymnus* addresses itself to God. Cardano wrote it at a time of great mental anguish—while his son was on trial. I have no doubt that the order is carefully thought out. It expresses a point of view much more critical of the Artistotelian system than that from which *De Subtilitate* was written. Accordingly, the mathematical writings rather than those on logic are at the head of the list. This seems to suggest that one cannot properly understand Cardano's philosophy unless one has been trained in mathematics. Of course, this is of programmatic significance only, since Cardano was still a long way away from a description of nature in strictly mathematical terms. He was, nonetheless, a mathematician of importance.

Toward the end of *De Libris propriis,* Cardano offers some basic suggestions as to how books should be written, adding that "anyone who will heed this can also easily improve my books." He says:

First, as far as style is concerned, one must not adhere too closely to Ciceronian rules. If it is necessary to talk of "entrails" one ought to do so, especially since Cicero never mentions them. The entrails of the cow are actually nourishing and when properly cut up, seasoned and cooked, quite tasty as well. It is enough that Horace and Pliny have lived. On the other hand we should not take the liberty of using Greek terms perhaps clumsily when Latin ones can serve the purpose. We should also avoid belching up words like "Haecceitas,"[32] and should not use foreign words whose meanings often are unclear. The word should suit the thing described. When talking about architecture, we should not use the language of Livius, nor should we try to talk about medicine in the style of Cicero's legal orations. It is better to use consistent style even if it is not a generally accepted one, rather than using a mixture of styles for the sake of catching a pretty turn of phrase here and there. One should avoid an unduly elevated style, without going to the other extreme, stumbling this way and that like someone about to lose his footing. One should make an effort to avoid vagueness, which all too often creeps in anyway. In the course of time, archaisms, basic inadequacies, and normal changes in language will create obscurities of expression that will render books useless. If one stays close to idiomatic usage, the writing will be clearer and more easily comprehensible. Use everyday language, and this will ensure that everyone can understand you. Do not, however, become trivial—this is a sign of superficiality. Avoid verbosity because it wastes time, our most precious good. Pay attention to the things Galen was criticized for, and don't think that they will bring you praise. Always remember: Words are there to describe things, not things to illustrate words. Although a lad wouldn't, one can learn more about the author's experiences in a slender volume by Horace than in the many books of Ovid. On the other hand, don't be too brief, because this easily leads to obscurity and abruptness such as one encounters with poets and also with Hippocrates. But then the latter had some justification for this as he had to write on wooden tablets. Don't use archaic expressions that first have to be explained, although they may well have been common in the time of Cicero. Matters concerning contemporary society which are at present familiar to everyone should not just be mentioned in passing. This is how so many things of our historical past have been obscured. Matters that are common knowledge in Italy today will be unknown to the Mamarians and Paropasmians a thousand years hence. The fate of literature everywhere is to flourish and then perish.

In this passage, Cardano expresses above all his criticism of "Ciceronianism," thus taking the side of Erasmus in a then-famous dispute.[33]

Cardano's Latin is highly individualistic and perfectly reflects his somber, brooding personality. He has a pronounced tendency to digress, although he does not seem to be aware of it. This may be explained by the fact that he sees a unity of things that is beyond logic. Sometimes his statements are puzzlingly brief. In those instances where he suddenly drops a train of thought and then picks it up again later, or where he refers to something he has not yet explained or has mentioned much earlier in passing, the meaning and coherence of his thoughts is often rather difficult to determine. Yet we must credit him for his endeavours to use words "to describe things," because he does indeed have something significant to say. However, he followed his own excellent rules on effective writing to only a very limited degree. It is perhaps this that prompted him to recommend them to those who might want to improve his works! Many of his writings seem to be direct notations of his thoughts as they came to him, and often he did not bother editing what he had written.

2 *Cardano the Physician*

CARDANO WAS A PRACTICING PHYSICIAN as well as a professor of medicine. Selections from the lectures he gave at Pavia and Bologna are included in his published works. They take the form of detailed commentaries on the classical medical authorities, Hippocrates, Galen, and Avicenna.[1] These commentaries constitute a large part of Cardano's medical writings and supply noteworthy information about teaching methods in the universities at that time. In addition, Cardano wrote numerous treatises on special topics, such as *De Dentibus* and *De Urinis,* as well as introductions to practical pharmacology like the *Ars curandi parva.* The latter text begins with a brief outline of his theory of physiology, which is essentially the same as that of Galen.

Cardano's significance as a doctor lies, however, not in his theoretical views but in his abilities as a medical practitioner. It is to this that he owed his worldwide reputation. A collection of fifty-seven "Consilia medica"[2] included in his published works gives us an idea of the manner in which he attended to his patients. Number twenty-two concerns the Scottish Archbishop Hamilton, and number fifty-two lists the recommendations concerning diet and mode of living which Cardano personally handed to the patient: "Ephemeris seu vitae ratio pro Reverendissimo D. Archiepiscopo Andreae DD. Joanne Amulton." While all the other consilia only fill between one and five pages in folio, number twenty-five fills thirty. It is obviously the most important case in the collection. The consultation report includes extensive theoretical deliberations as well as numerous formulae for remedies to be considered—although very few are actually used. In addition to these two consilia I should like to discuss number nine, which concerns a case of "hypochrondriac melancholia." It will impress upon the reader the sensible human attitude that was characteristic of Cardano's practice of medicine.

Numerous statements contained in these consilia are only comprehensible if one is familiar with the theory of physiology prevalent at the time. It is the Galenic theory, and it continued to exert a decisive influence on medical thinking into the eighteenth century,[3] although its foundations had already been shaken in the seventeenth century by Harvey's discovery of blood circulation (*De mortu cordis*, London 1628).

In Galenic physiology major importance is placed upon the four humors. They each originate in a particular organ, or rather, they are secreted by this organ from the blood. These four humors are: the blood that is produced by the nutrients in the liver, the yellow bile from the gall bladder, the black bile (melancholia) from the spleen,[4] and the phlegm from the brain. Coordinate to the four organs, or rather to the respective fluids, are four basic qualities: moist-warm to blood, moist-cold to phlegm, dry-warm to yellow bile, and dry-cold to melancholia (black bile). The harmonious mixture of the humors and basic qualities was called "temperature," with some variations according to "temperament." If, for example, phlegm is the naturally predominant humor in a person, he will be phlegmatic and of a cool-moist constitution. If this temperature changes, a person falls ill, either because the humoral system has become unbalanced, or because it does not suit the person's constitution.

The blood also contains a "spiritus," which Cardano describes as an "ethereal body."[5] As *spiritus naturalis,* it flows together with the venous blood from the liver to the heart. In the heart, which is the source of the body's heat, it comes into contact with air and the "soul substance" contained in it: the *pneuma,* whereby it is changed into the *spiritus vitalis,* the vital spirit. The blood now becomes arterial blood, the heart is stimulated, its beat strengthened, and the palpitations communicated to the arteries. In this way the vital spirit warms the body. It would be dangerous, however, if during this process the heart became overheated. Its temperature is kept down by respiration, which also cools the brain—an organ which is naturally cool.

As the arterial blood flows from the heart to the brain, the *spiritus vitalis* becomes the *spiritus animalis*—the *pneuma* which then fills the ventricals of the brain and the nerves. According to Galen, this transformation occurs in some mysterious organ at the base of the cerebrum, the *nexus mirabilis* or *rete mirabilis.* Cardano states that this famous object had been found in cattle, but that there was no evidence of its presence in the human body.[6] Andreas Vesalius, whom Cardano regarded as by far the most estimable physician among his contemporaries, had not found it either, but had nevertheless described it (*Fabrica humani corporis,* Basel 1543).[7]

It was thought important that in order to maintain the right "temperature," the fluids—in particular the phlegm and the two biles—must be continuously excreted, or they would become viscous and stringy, begin to decay and become harmful. If, for example, phlegm cannot be discharged because the nasal passage is obstructed by a cold, congestion and numbness in the head will result. Furthermore, the brain will not be sufficiently cooled, which will apparently cause fever.

These ideas do not, of course, come close to our modern understanding of physiology; yet they are concrete and perhaps describe everyday subjective experiences more clearly than do modern scientific explanations.

In its psychological connotation the theory still lives on today. We speak, for example, of the temperament of a person and say that he is melancholic or phlegmatic; we even refer to a person's sense of humor without in either case giving a thought to the phrase's origin in the old humoral pathology.

I think that the continued success and practical value of this theory over the centuries is due at least in part to that portion of it which is expressed in this linguistic usage. Many illnesses are caused by physiological as well as psychological factors. In such instances psychological problems manifest themselves physiologically; they cause physiological dysfunctions and may in time cause organic changes. Today we speak of psychosomatic illnesses. The old physiology reflects subjective perceptions, and with its notions of vital spirits and opposite qualities it is excellently suited to represent psychological problems symbolically. Thus, in those days it could indeed be of great value in the treatment of patients. This was especially true then because the physician had very few remedies at his disposal which we would consider effective. Cures that did work were generally non-specific. But we might keep in mind that even today we often do not know exactly how a certain medicine works and what its effects will be.

In the case of acute infectious disease, however, the physicians of the sixteenth century were practically helpless, and a doctor had the right to refuse treatment. On the other hand, severe chronic illnesses were often treated successfully by dietetic prescriptions and by measures of a psychological nature. In this, Cardano apparently excelled.

In the twenty-second consilium (the one for Archbishop Hamilton), Cardano remarks by way of introduction that the physician ought to know the patient in good health, since illness is a deviation from this state. The more a patient deviates from his normal state, the more severe the illness actually is. Since Cardano did not know the archbishop, he had to use the fiction of standard normalcy as the basis for his diagnosis.

He did, however, have the report of the archbishop's personal physician, Dr. Guglielmus Casanatus. From it he learned that the bishop, now in his fortieth year, had for the past ten years been suffering from periodic attacks of asthma. He contracted it while out in the sun during the dog days. His brain suddenly became overheated, the elevated temperature caused a chill, and a discharge of phlegm from the brain brought on the first attack. The hoarseness associated with it was treated successfully, but there remained a cold-moist disorder of the brain. As a consequence of such a disorder, an abnormal substance will gradually accumulate in the brain which cannot be discharged properly. Periodically, a flow of the harmful fluid gets into the lungs, causing another attack. The attacks usually occur during full, half, or new moon. In the course of the attack the patient expectorates a large amount of watery, almost odorless mucus, which later on thickens until it can no longer be brought up even by the most violent coughing. This causes the asthmatic condition and the accompanying snoring sound. The patient does not get enough air, and his chest expands greatly. His general constitution is weak. During the attack his face looks ashen and sunken (*facies Hippocratia*), but afterwards it hardly looks different from that of a healthy person.

After Cardano had gotten to know the bishop's physician, he questioned him most closely, and he learned the following: the attacks did not only occur during the changes of the moon, but more frequently. Then again, the patient might be free from an attack for two or three weeks if he adhered to the proper daily regimen. (An attack usually does not last longer than twenty-four hours, although longer attacks do occur.) The bishop liked to sleep a lot, but because of his many duties often got too little sleep. He enjoyed eating and drinking, his bowel movements were light, and he perspired easily. In view of this information, Cardano felt that the patient would well tolerate a fast. Cardano also learned that the bishop was an irascible man.

Having gathered these facts and having witnessed an attack, Cardano understood what caused the illness. He disagreed with Casanatus on a number of points. For instance, Cardano believed that since the discharged mucus is initially thin and watery, it could not have been accumulating in the brain over an extended period of time. Had it remained there for something like eight days, it would certainly have started to decay and become foul-smelling. Over a period of ten years, this would inevitably have caused some damage to the brain. This did not, however, happen; the bishop's mental health was unimpaired. Furthermore, the expectorations come not only from the head but also from the general area of the thoracic cavity. This is caused by poor metabolism (*concoctio*) in the stomach, the liver, and the limbs, and is responsible for the patient's emaciated condition. After further theoretical considerations, Cardano concluded that the imbalance in the brain was not of a cold but of a warm-moist nature. On the basis of this diagnosis and in view of the bishop's lifestyle, Cardano prescribed the therapy. It is outlined in detail in the fifty-second consilium.

Here, Cardano gives the bishop two reasons why a cure had not been effected before:

> One reason is the long duration of the illness which has weakened the body. Meanwhile, your Excellency has relied entirely on medical measures for help instead of freeing yourself from the heavy burdens of your office. The other reason is this: because the illness is of a moist nature, remedies of a dry quality have been used. But Galen teaches (*De tuenda sanitate*, book V, chapter 5) that old people—and the physical condition of the patient is that of an old man—who suffer from a heavy discharge of mucus will initially experience relief from dry medicines, but that with prolonged treatment they will incur irreparable damage.

Therefore, Cardano recommends the use of moist remedies:

> It is wrong to treat the patient with forceful laxatives, to insist on a diet of dry foods eliminating all liquids from it, although this does seem more appropriate than the treatment described below, which by necessity at first seems to have a harmful rather than a beneficial effect.

Most important is a sufficient amount of sleep, approximately nine hours. The patient should awaken refreshed and invigorated. The bishop is advised to seek long periods of sleep and to rest if he is unable to sleep. "It is not enough to know what is essential, one has to know the most essential." Only if the patient gets enough sleep can he hope for a cure. All this is most emphatically impressed upon him.

Upon awakening in the morning at six o'clock, the whole body—beginning with the legs—should be gently rubbed down.[8] This should be followed by an application of almond oil to which a small amount of pitch has been added. After the rubdown, he should move his bowels, preferably without inducements. If this is not possible, a mild enema is to be prepared by boiling down dog's mercury (*Mercurialis* L.) or strawberry blite (*Blitum* L.) and adding some honey and a pinch of salt to the extract. The enema should not be an irritant. All this should be accomplished by eight o'clock. For his first breakfast he should then take a glass of lukewarm, lightly sugared tea (Althea off.).[9] After that he should comb his hair with an ivory comb.

Outdoor physical exercise should be next. Archery and hiking are recommended, as is reading aloud—but only in a soft voice; this is very important.

Once a week around eight o'clock—after having moved his bowels and before exercising—his head should be washed with very warm water, immediately followed by cold water in several successive jets. He will not notice any positive effects of this procedure at first; he will even find these spouts most unpleasant, since he is, after all, not accustomed to having his head washed. But in time it will strengthen his brain. This should be done only once a week. After these washes he ought to stay indoors for a while. He should always chew on a bit of pistachio resin (*Mastix*) when he goes out, and also during his exercises. This helps to eliminate the moisture from the brain as a great amount of saliva is being secreted. Breakfast should be at nine o'clock, consisting of bread moistened with chicken broth, some roasted chicken and small quantities of light white wine, never more than half a pint—less would be better. If he is thirsty, he can rinse his mouth with wine; drinking is a matter of habit. After the meal, until noon, he should avoid any official business; instead, he should relax with light conversation or listen to music. Thereafter, he may attend to his affairs for two hours, but not do any writing himself. Whenever possible, he should be spared troublesome news. At four o'clock, he should go horseback riding for an hour, with good company that stimulates conversation. He may receive visitors until dinner at seven, but only while seated. Dinner should be lighter than breakfast, and he should also drink less. A spoon of honey taken before the meal aids the digestion, relieves the lungs, and tastes pleasant besides.

Frequently, the evening meal should consist only of fresh goat's milk and bread. He should retire shortly after eight o'clock. However, he must not sleep on eiderdown,[10] but on quilts filled with silk. The pillow, with a cover of linen or better still of hemp, should be filled with corn chaff. If this is too hard, dried seaweed can be used. The air in the bedroom should be cool, with the window left open until midnight.[11] A wood fire should burn in the bedroom during the day, but must be extinguished at night. The acquisition of a large reliable clock of the kind used today by all princes in Italy will ensure that the regimen will be closely adhered to. In winter, instead of going horseback riding, he might have someone read to him, or he might play a game, but without getting irritated. Nothing is more harmful than being irritated.

If he has an attack, he should not eat anything, except take a little honey and some clear chicken broth. (A chicken is boiled in a gallon of unsalted water until the water has reduced to six pints.) When the attack is over, he should eat an egg yolk with a little ginger and some bread dipped in broth; he should drink no wine. Whenever he feels the onset of an attack, he should immediately induce vomiting by putting a quill in his throat; this will give most effective relief. Then, the arms and legs should be bound with soft leather, after a quarter of an hour be unbound, and the procedure repeated.[12] These are the most effective remedies, and they form the core of the treatment. When the illness subsides, ample sleep is the best thing. The windows should stay open, but the fire should not be lit.

This is a moist-cool treatment of a warm disturbance of the body and the brain. As Cardano himself points out, the therapy takes into account things such as the patient's diet, his bedding, ventilation, even his emotional state. The regimen prescribed is quite obviously a strict one, ensuring that the patient does not expend an inordinate amount of energy on his breathtaking political office. Apparently, this course of treatment produced amazingly rapid improvement.

Surprisingly, Cardano prescribed almost no medicines. In the consilium, many are considered and the recipes given, but in the end Cardano relies primarily on mallow-root tea.

The dietary recommendations for the archbishop are, in many respects, nonspecific. They basically conform to Cardano's idea of a healthy life. This is particularly apparent in the ninth consilium, which deals with a case of a patient suffering from hypochondriacal melancholia. In this instance, Cardano did not see the patient, but received a long letter from him in which he implores

Cardano for advice. Since the diagnosis is based on that letter, Cardano transmits it in its entirety, and I shall do the same. The patient writes:

1. A certain hardness is located on the left side beneath the "false ribs." It is hardly visible, but it is sensitive to touch and pressure. It remained from a fever I had five years ago. Some say it is a muscular induration, others think it is sclerosis of the spleen which has been blocked by thick mucus. This tumor does not feed on the surrounding tissues; it also does not cause any pain. It is instead nourished from the epigastric region, causing contraction of the diaphragm. This tension seems to spread to the chest.
2. I am actually feeling a tightening of the chest, and I have great difficulty breathing; my lips are dry and breathing is accompanied by an internal discharge of mucus. Exhaustion of the vital spirits and pressure on the trachea frequently impede my speech, and I have great difficulty in regaining it. I secrete stringy, viscous saliva. It seems as if a catarrh causes distillation into the lungs and is thus interfering with breathing.
3. Furthermore, I feel strong heart palpitations which cause a high pulse rate in the arteries as well as a severe headache. This is most pronounced when I exert myself physically, or when I experience great anger or other strong emotions; they seem to suffocate me.
4. My head is weak, and so are all the inner senses, in particular my memory. Moreover, I feel a dulling of the mind, a dimness, and an apathy that turns to stupor whenever I want to study.
5. When I do study, I often have a ringing in the ears, and I also feel dizzy.
6. Due to poor circulation, the tips of my fingers often turn white. After a while, they become numb, and only rubbing or some similar means can restore the flow of blood.
7. I frequently feel more unhappy than I would like to be.

My anger, and other emotions as well, tend to be more intense than they ought to be. For this reason, I am much afraid of drafts.[13]

8. Because of my weak constitution I have been taking numerous drugs since my earliest years. Over the course of time this must have weakened my physical strength. Sir, I beg you to prescribe something that will save and prolong my life. Let me know also in which way, how often, and at what times I should take the medicine.
9. In addition, I believe that my blood is dried up and impure. The spots which have recently appeared on my cheeks are an indication of this.

Finally, I implore you to let me know your judgment of my illness: investigate the causes and try to find a cure. Should the fee I have offered seem inadequate for your efforts, I will gladly and generously increase it. I would in no way want to appear ungrateful to you.

To this Cardano replied:

If this description is accurate, the patient's condition is undoubtedly caused by poor digestion in the stomach and liver, together with a heavy discharge of mucus from the brain. This mucus gets into the lungs. Because of the high temperature in the liver, because of the patient's age, and perhaps also because of the condition of his heart, the puss turns into black bile, causing the enlargement of the spleen. This is also the cause of his delusions; he is really not as slow-witted as he believes he is, this notion is fostered by melancholia. The form of his letter is proof of this: everything is well-reasoned, his expressions are vivid and to the point, he presents his request in an intelligent and skillful manner, demonstrating thereby that his mind and memory are strong, that he is intelligent and that he is, as I said, suffering from windy melancholia (*melancholia flatuosa*) and thick blood. This is basically his condition. I would say—although one might draw many more conclusions—that he is cautious and sensible, but that his condition puts him into a constant state of anxiety. This not only destroys any feeling of happiness, it also destroys his reasoning powers. Consequently, he imagines that he is going to meet with some great misfortune. Although he is most certainly ill, the situation is actually not all that critical. Not one fourth of the things he is imagining is going to happen. I am not certain whether his is a case of some deep but passing sorrow or of some enduring sadness which is, however, not very deep. At any rate, he must first of all eliminate the reason for his illness. He should live cheerfully and be mindful of Democritus's saying: "since our life is short nothing of great importance can happen to us." He will learn from experience that nothing is more beneficial than a cheerful outlook on life. This is what I call a philosophical attitude: a long way, serious care—for care dispels cares—and then play and merriment. Medicines will be helpful, so will certain foods, and in this case some gratifying reading matter as suits personal taste. It is difficult to cure the mind without using the mind. Take this to heart: nothing is more beneficial than moderation. Moderation is the veritable "aurum potabile," as Aloaesius Cornarus[14] so aptly put it. A person who tries to expel the harmful fluids from his body without changing his basic habits—this means eating and drinking in moderation—goes about it in the wrong way. He may alter the disease, but he will not eliminate it. However, once the body has been purified, will there still be any need for medicines which will only harm him? Why not restore health by moderate exercise of body and mind, and by adhering to judicious dietary principles? A scholar in particular will benefit from this. There is empirical evidence that the most sagacious men preserved their health into old age by these measures. If every sick person kept this in mind, much unnecessary suffering could be avoided. There are really only four kinds of people who will take recourse to medicines: first, those who have been persuaded by doctors eager for pecuniary rewards that medicines are necessary, convenient and cheap; second, persons who are disdainful of leisure and pleasure and instead seek excessive work, thereby neglecting what is useful and pursuing what is harmful; third, people who long ago contracted such a severe illness as the lues that they must by all means try to rid themselves of it; finally, persons whose illness is the result of some form of violence such as a wound, chaining, spraining, pestilence or poisoning, in which case suffering will be acute. It would, therefore, be foolish to wait for the slow healing process ensured by a moderate life. Precisely that, however, shall be the only aid for you to preserve your life; if you do not observe it you cannot hope to recover. And think what many other benefits are to be derived from a life of moderation! True wisdom is the result of temperance and it fosters our common sense, memory, joy, peace of mind, and longevity. To this, then, you should pay foremost attention. If you do, then the medical remedies will certainly bring about a cure within a year. (For medicines are needed where the length of the illness or the anxiety of the patient[15] makes this advisable, or when we can thereby shorten the period of recovery.) If you believe, however, that you could get along with the help of medicines alone, without a well regulated life, then a cure is uncertain. In that case I cannot promise anything, and I am not sure that I could be of any help.

I am now going to discuss the matter in detail. Two meals a day, each consisting of bread dipped in meat broth with raisins or almonds or sweet almond biscuits. If this is insufficient, you may eat one small bird. Drink white wine mixed with plenty of water. Occasionally, eat some leeks or raisins boiled in meat broth. The other dishes shall be capers or chicory cooked in vinegar. Meat should be from castrated animals—from lambs or young pigs—which have been kept in the field rather than in pens. Of the large birds which should, however, have small beaks, indian chicken and pheasant are best. The best part of chickens and capons are the wings and the liver. The bread should be very light, fresh, well-baked and well-fermented. The wine should always be diluted, in the evening as well as in the morning. Everything else is prohibited, except for scaly, broad, light fresh fish such as perch. No root vegetables, no milk dishes or custards, or other kinds of fish. This is the way you should live, and remember: food and drink are mere physical sustenance; beyond that you should be indifferent to them. You may eat fresh fruit, such as the first cherries, plums, figs, peaches or grapes, but eat only one kind at a time, and this always before meals. Even when you feel well, do not keep yourself awake, study or engage in sexual activities. Do not exert yourself. All foods should be well cooked. Eat with care; avoid a great variety of foods; be satisfied with one kind, or at the most choose two. The most important question is that of proper quantities. The answer is difficult. Everybody is looking for it, but nobody can find it, because nobody can define moderation, although everyone praises it. The meager fare recommended by Cornarus will not be to everyone's liking, and it will not be appropriate for everyone, since differences in age, sex, constitution and physical build have to be taken into account. One must also consider how much energy a person expends, what his habits are, and what kind of environment he lives in. It will not suffice to follow the "princep's" advice to leave the table when you still feel a little hungry. We have found that this is not only unnecessary but in most cases utterly useless. Let us take then as a criterion for proper measure that a person after having finished his meal should still enjoy going for a walk, or sitting down to some writing. After seven hours the stomach should feel empty, he should not belch, nor have bad breath, nor feel any heaviness of the stomach at all. His sleep should be peaceful and uninterrupted, and when he rises at dawn he should feel no heaviness in the head. He should not want to lie down again, but should instead wish to get up quickly and go about his tasks. Let this be the measure of proper quantities of food and drink.

Whenever possible eat some rue (*Ruta graveolens* L.) before breakfast, seasoned with a little salt, oil and vinegar. This thins the black bile, liquifies the viscous fluids, loosens the flatus, improves the digestion, opens up blockages, counteracts toxins, kills worms, eliminates the sexual drive (*veneris prurigo*), strengthens the head, sharpens the mind and preserves good vision. He should also note that nothing is better for him than good air, air that is pure, dry and of an easterly direction. Physical exercise should be a pleasure. It should be done before meals, outdoors, inside only in bad weather. He should wear light clothing, but avoid drafts. He should cover himself well at night. In winter as well as in the morning the bedroom window should be closed; in fine weather it should be left open.

Cardano then discusses medical remedies. Because the patient has been poisoned, so to speak, by some thick cold substance, he will have to be gradually cleansed. Cardano mentions the fact that very effective drugs containing antimony are being used in Germany for this purpose, but he adds, "although I am not disputing that he could be cured by such means, the prescription of such a treatment conflicts with my overall philosophy. I do not use these substances because they are unnatural and I believe them to be dangerous."

Cardano prefers the use of the hellebore root (*helleborus niger*), and he extols its special merits.[16] He knows, however, that *helleborus niger,* of which the roots are used for the medicine, is highly poisonous. For this reason, a preliminary treatment is required, in which a syrup is taken internally and an ointment applied externally. The composition of these medicines is exactly specified. To give the reader an idea of the state of medical science at that time I am going to quote here the most important recipe—that for *helleborus niger.* The weights used are the apothecaries' pound (ca. 370 grams) and its units:

1 pound = 12 ounces (oz)
1 ounce = 8 drams (dr)
1 dram = 3 scruples

Recipe: 1.5 dr roots of *helleborus niger,* the variety from the mountains bearing red flowers. Cut the roots and insert the pieces into cuts made in a large beetroot. Place the beetroot in three ounces of pure honey and let this stand for twelve hours in a warm place. Then pour off the liquid honey and carefully squeeze the beetroot into it. Discard beetroot and hellebore roots. Mix the honey with the following broth (or take the honey by spoonfuls first and then drink the broth): Place a hen in twenty pounds of water. Add:

- 1 dram each of polypodium (fern) and senna leaves,
- 2 drams each of the greens and roots of leeks, sprouts of hops, blossoms of fumitory (fumaria, off.), fresh or dried, fresh leaves of balm (mint), St. John's wart, and marjoram.
- 2 ounces of violet leaves, 2 drams rosemary-root,
- 1 ounce black myrobalan (Emblica off.),
- 5 drams whole peppercorns

All this should boil until it has been reduced to two pounds.

Six ounces of the broth should be taken very hot. Do not take the medicine the following day, but every day thereafter until the broth has been used up. (The honey must be prepared daily; the broth will be enough for four days.)

This principal treatment is succeeded by a follow-up treatment using less powerful medicines. Finally, Cardano gives the recipe for his "miracle-powder," guaranteed to strengthen the mind, open up blockages, purify the fluids that cause melancholia and be very beneficial for the heart as well. This powder is composed of small quantities of golddust and ground precious stones and of a larger quantity of ground lodestone. It is eaten mixed with sugar. After having eaten this, no more food is taken for seven hours and one should go for a walk to loosen the bowels. "You will find it most effective."

The final question raised in the consilium is that of bloodletting, a question not to be dealt with lightly. Cardano feels that in this case bloodletting is unnecessary because the illness is of a cold quality, of long duration, and because it is producing large amounts of black bile. Cardano intimates that there are various reasons why physicians, who are essentially ignorant of proper treatment, make much use of this procedure. It gives the impression that for once they are doing something important. And they are actually following Galen, who insisted that bloodletting is always beneficial. Last but not least, they like to employ this measure because it allows them to prescribe a less strict diet, and this will please particularly their younger patients. In the case at hand, bloodletting is considered absolutely unnecessary unless the patient should decide not to follow the prescribed diet and neglect the much needed phsyical exercise. Then a measure which is inferior, but which can at least be empirically credited with some usefulness, will be better than nothing at all.

Almost all the herbs that Cardano prescribed for use in his medicines were until recently designated as medicinal. I am not in a position to judge how effective or ineffective these remedies are. The miracle-powder is obviously a placebo; yet it is well-known that such means often produce extraordinary results. It is clear that this patient requires treatment beyond the dietary regimen which Cardano regards as crucial. Something must be done about his states of anxiety. Only then will he be receptive to Cardano's ingenious psychological advice. The latter is really at the heart of Cardano's counsel, as evidenced by the change of address from the more impersonal third person to the direct form of address. Cardano does not separate either in theory or in practice the psychological and the physiological factors, which seems fairly characteristic of medical practice at that time. Cardano's diagnosis points to a pathological imbalance of the humors, and he defines the illness as a metabolic disorder—that is to say, poor digestion (concoctio) in the stomach and the liver. In addition to this, he has diagnosed a psychological disorder evidenced by the profuse discharge of mucus from the brain. Together, these imbalances produce the black gall which in turn causes the patient's delusions. He suffers from a form of depression accompanied by very unpleasant physical symptoms.[17] Because the illness manifests itself physically, it also has to be dealt with in physiological terms, although only an appropriate regimen that will take both the psychological and the physiological factors into account will assure improvement. Looked at in this way, there is even today very little one could object to with regard to Cardano's point of view and procedure.

The separation of physiology and psychology—one of the consequences of Cartesian philosophy with its sharp distinction between mind and matter—led to new developments in medical science, but it also raised new problems.

Cardano lived one hundred years before Descartes. His philosophy is based upon the concept of unity, but it does lack "clarity and distinctness," as the examination of his writings on natural philosophy will show.

In his medical practice, however, his philosophy greatly contributed to the successful treatment of his patients. His impressive personality assured him of the authority to convince his patients that they themselves could and should do the most to ensure good health; this made him a great physician.

3 *Natural Philosophy and Theology*

CARDANO ATTEMPTS to conceive of the world as a unified whole. In accordance with the idea of unity of the terrestrial and the celestial, of the physical and the spiritual world he believes in a single vital principle: the "World-soul." At the same time he is greatly impressed by the profusion of phenomena that he perceives in the world and that he wishes to include in his extensive knowledge. It seems to him that a single principle cannot account for the wealth of diverse forms of which he is continually aware. Instead, many principles must exist. Clearly, the old dialectic of "the one and the many" is very much part of his thinking. Anyone studying Cardano's philosophy of nature ought to read *De Natura* first, as Cardano himself advises in the list of his works, *De libris propriis*. Closely related to *De Natura* is *De Uno*. Both works were probably written around 1560, but *De Natura* was not printed until the publication of Spon's edition of the *Opera*.

De Uno[1] is a brief exposition of Cardano's concept of unity. It shows that this idea is fundamental to his philosophy, hence the inclusion of this text among his writings on logic. *De Natura*[2] is not very extensive either, but it still fills fifteen double-column pages in folio, the equivalent of about sixty pages of modern typography. In *De Natura* Cardano discusses in great detail Aristotle's concept of nature and juxtaposes it to his own opinions. I would like to leave open the question of whether he already differed with Aristotle when he wrote *De Subtilitate* in about 1550, or whether he only became decidedly antiaristotelian around 1560, as is evidenced by *De Natura*. One can certainly view *De Natura* as an introduction to *De Subtilitate*, as indeed Cardano himself conceived it.

Anyone expecting to find a clear and organized train of thought in these writings will be disappointed. One of the reasons for this is Cardano's way of presenting his ideas as they come to him. He begins with considerations and reflections which seem plausible to him or which were considered plausible at the time he was writing. In the course of further reflection he often realizes that he must correct his initial statements, or that other hypotheses are more in accordance with his ideas. However, he does not retract what he said previously, but instead continues the argument from his new position. He apparently wants to describe the actual progression of his thinking rather than present a train of thought after it has been rationalized. Today, of course, ideas are presented in rationalized form. Only after reaching the end of one's deliberations does one realize the entire course one's argument ought to have taken in order to have smoothly arrived at the insight which was achieved. It is this logical path which is presented to the reader. But Cardano does not write like this; even his philosophical writings are, as it were, biographical in nature.

Often one gets the impression that Cardano pursues several distinct but correlated lines of thought at once. Since all is unity all shall be grasped at once. This unity is brought about by various correlations which are rather difficult to present discursively, as they would require involved explanations. And then the whole, which is to be encompassed in a single view, might easily be lost. For this reason, Cardano often resorts to pointed intimations which give his statements an aphoristic quality. The overall impression is indeed lively, but also somewhat desultory, confusing, and obscure.

The great success Cardano had as a writer shows, I think, that his contemporaries took little note of such shortcomings; the time of Cartesian "clarté" was yet to come. My attempt to present Cardano's natural philosophy in a somewhat systematic manner is therefore also an interpretation of his thoughts. What I describe as a deduction from Cardano's presuppositions is for him frequently an empirical point of departure for his reflections, since he alternates inductive and deductive arguments in a rather peculiar fashion.

The point of departure for his reflections in *De Uno* is the proposition: "The One is good, the Many are bad." Therefore, good is what is unified, and unity is brought about by order. Hence, the One is a principle which institutes order. And since order can only exist as an ordered plurality, the multitude, although bad in itself, is nevertheless necessary. Such ordered multiplicity reflects divine love, goodness and wisdom. From this various inferences can be made. As multiplicity exists for the sake of order, which stems from unity, the Many also originate in the One, since a plurality cannot of its own produce unity.

With regard to humanity, this would be interpreted as follows: A single human being is good and many human beings are indeed a multiplicity of good, but certainly not better than a single human being. However, if many people unite in civic or professional organizations, each one is brought into relation with a unity. Such an association is therefore better than many independent individuals, better even than one single human being. In terms of political science this would mean that a monarchy is better than an aristocracy, which in turn is better than a republic. This is the Aristotelian viewpoint.

Since that which is good is also beautiful, it follows that in music consonance is more beautiful than dissonance, since in consonance the sounds form a harmonious whole.

Similarly, uniform ordered composition is the origin of beauty in the visual arts. Therefore, the products of one single artist, even if he is not particularly significant, are more beautiful than those created by a group of artists, even if each one of them is important in his own right.

Unity is always unity of something which must actually exist. Therefore, unity is also actuality, whereas the merely potential belongs to multiplicity. But since multiplicity is bad, mere potentiality—the "materia prima" of the Peripatetics—is extremely bad and therefore does not exist.

If one wants to evaluate these viewpoints in terms of natural philosophy, if one wants to comprehend the order of the universe and its unity—since order produces unity—one must not harbor the false notion that entities exist for the sake of each other. Specifically, the world was not created for man, but rather, all things were created for the cosmos.

From the One emanate chains of order—there are many—but as they all lead back to the One, they form an organized whole. The individual entities do not appear ordered if one looks for order in their correlations. These entities can belong to totally different chains of order whose unity can only be discerned by tracing the origin of each back to the One. To do so for the whole world is impossible. After all, who could understand the world and every single one of its orders? One can, however, form at least an idea of such chains of order in the world by looking at man, as he is a microcosm.

The unity of man is a unity in principle. Each particular human being is a singular individual who thinks, feels, digests, walks—who has an immortal soul. This is the eternal principle of man which governs the body and the vital spirits. There are various parts to a human being's body—hands and feet, for instance. Hands exist to seize things, feet to walk, but the hands do not exist for the sake of the feet. On the other hand, the throat exists because of the stomach, the mouth because of the throat, the teeth because of the mouth, the lips because of the teeth. These organs, then, form a single order without there being a direct relationship to the feet, for example. The soul unifies all of this, but it is itself not part of the body. We feel different things with our hands than with our feet, while some organs do not feel anything at all. Nerves, for instance, do not feel that they are being nourished. From such reflections Cardano concludes how one ought to perceive the essence of the soul. He says:

> The soul, which is the unifying principle of the human being, is neither a continuous extension nor a multipartite composition. It is itself a One. Therefore, it exists by itself, and it is neither in space nor in time, since it would then be spatially or temporally extensive. Therefore, it is not in the body—it is nowhere. But from it life streams into the body, in Cardano's words the life "through which we accomplish everything that we do, or through which we, as sufferers, become perfect."

Life is not a principle, and it is present in every part of the body.

Just as the soul functions in us, a soul also operates in the universe. It, too, is imperishable and is nowhere. As is the case with man, the soul also produces the sympathy and antipathy of all things in the cosmos.

Inserted into these reflections is a more logical analysis of the various ways in which the concept of unity can be understood. Cardano remarks that people often speak of "numerical unity." But this does not really make sense. "Number is a fiction of the human mind. Its principle is the One. 'Numerical unity' is therefore a fiction of the soul." Cardano first lists eight ways in which one can speak of unity:

1. God is truly one, for he cannot be conceived of as divisible.
2. Heaven is one and unique, but it can conceivably be divided.
3. Mass is an agglomeration of similar but independent parts; it is discontinuous.
4. A wall is a continuous body, but it is not composed of uniform substance.
5. Milk is continuous in form and uniform in substance.
6. The color white is a uniform accidental phenomenon.
7. A species is a logical unity.
8. Man is a unity which has a soul.

It should be noted that in this sequence "milk" as a continuous and uniform substance is cited after "mass" and "wall." Evidently, Cardano is thinking of a particular mass and of a particular wall, whereas milk refers to milk in general, that is to say to a concept, which then leads to "white" and "species."

Man, as a microcosm, is mentioned last, giving the list a certain completeness. But Cardano does not stop there. Man calls to his mind the question of order and of fate. These are two different kinds of unity: fate requires time—it cannot be conceived of outside of time; but this does not apply to order. Cardano concludes these reflections with the words: "Thus we have already found ten kinds of unity," obviously implying that there could be more.

It seems to me that the essential purpose of these inquiries is to show that the very concept of unity contains the idea of multiplicity. This gives logical backing to the idea that unity is the origin of plurality.

In *De Natura* Cardano presents his philosophy of nature. Again, the mode of presentation gives the impression of disorder. One of the reasons for this—in addition to those already mentioned—is that Cardano continuously engages in polemics against Aristotle,[3] in particular against his theories of growth and decay, of the elements and of nature, as stated in *De Ortu et interitu* and in book VII of the *Metaphysics*.

Although Cardano considers several of Aristotle's doctrines to be wrong, he nonetheless regards his mode of investigation as exemplary, both in terms of methodology as well as objective. He does, however, often arrive at totally different conclusions. But Aristotle remains his model, and he continues to employ Aristotelian definitions, although he has found them to be incorrect. In order to understand Cardano, then, it is necessary to recall the Aristotelian theory of growth and decay and his theory of the elements. Aristotle's writings certainly do not always present his teachings with the desirable degree of clarity. One of the reasons for this is that Aristotle also continually engaged in polemics against his predecessors such as Empedocles, Democritus, and Plato. Cardano does not fail to point out that it was much easier for Aristotle to refute the insufficiently substantiated doctrines of his predecessors than to construct something original. But in our context it is not important to know what Aristotle's actual opinions were, since for Cardano Aristotelian philosophy is synonymous with the philosophy of the Peripatetics as it was expounded by Alexandrian, Arabic, and Scholastic commentators.

According to these commentators the Aristotelian doctrine was as follows. There exists a substance which is totally inert and has no properties whatever: the "materia prima." Its nature is pure potentiality. Then there are the fundamental qualities which logically as well as actually have the nature of pairs of opposites: hot-cold and moist-dry. They occupy matter in such a way that only one member of a pair of opposites can be present in it at any given time. This follows from the logical nature of opposites, that is to say, from the principle of contradiction. This is how the four elements originate:

Earth is cold and dry,
Water is cold and moist,
Air is hot and moist,
Fire is hot and dry.

Following the pattern of eternal celestial motion, the elements undergo a cyclic metamorphosis: the earth becomes cold and changes into water, fire becomes cold and changes into earth.

Just as the elements change into each other as the opposites intermingle, all growth and decay comes to pass. The inner dynamics of opposites are a natural cause of all processes of nature. It lies, then, in the nature of things, in their contrasting qualities, that one thing emerges and another subsides. Thus, a healthy person falls ill, and the sick get well again. Health and sickness are opposites.

For nearly two thousand years, this theory was accepted as reasonable, and it also impressed Cardano. But he believes that it is based on incorrect assumptions; this has far-reaching consequences. Cardano states that privation—the absence of a quality—does not produce an actual opposite. A logical contrast such as hot-cold is not really a contrast. Blindness is not the opposite of sight, and death not the opposite of life. When the sun rises it brings us light. Yet darkness does not require a contrasting principle. It suffices that the illumination ceases. Cardano describes an actual contrast as follows: "An opposite is the nonexistence of something for the very reason that something else exists. But nothing means precisely nothing at all, because there is nothing existent."

From this follows that the element "earth," which, according to Aristotle, is cold and dry, has no properties at all. This would mean that the earth and the "materia prima" are identical. In actuality, however, earth is not at all the "materia prima," which is really a non-thing. Obviously, something which is supposed to be mere potentiality cannot actually exist. Everything which is existent exists in actuality. Therefore, it makes no sense to introduce a "materia prima" in addition to earth. It is earth, rather, which is the "heavy element."

Cardano observes furthermore that fire certainly cannot be an element. Elements must have a certain degree of autonomy and permanence. But fire needs constant nourishment or it will go out. It is nothing more than a heated fume or vapor. Since, according to Aristotle, fire is dry and hot, it would, in the absence of a "materia prima," be pure heat. As Cardano no longer regards fire as an element, heat becomes independent of the elements. Similarly, water becomes pure moisture. Water, on the other hand, does appear to be an element, and it ought, therefore, to have elemental characteristics. Cardano comes to the conclusion that moisture is associated with matter, whereas heat is not.

Heat obviously comes primarily from the sun, our most important source of warmth. Heat streams forth from the heavens—from the sun and probably also from the stars—down onto the earth, the sublunar material world. In this, Cardano concurs with the Stoic theory of celestial heat as transmitted to us in Cicero's *De Natura Deorum*.[4] The Stoics considered celestial heat to be the source of all life—a view we still hold today—and as the principle generating life was called "soul," they regarded heat as the "world-soul." Therefore, all living creatures are called "animals" because they have a soul ("anima").

To be sure, Cardano does not go as far as the Stoics. He by no means identifies heat with soul. But he does associate it with the soul, just as he associates moisture with matter.

As Cardano abandons the notion of the logical as well as actual dynamic nature of the Aristotelian opposites, he must replace Aristotle's theory of growth and decay with another. He believes that the soul and matter are fundamental principles of nature, with heat and moisture functioning as mere tools. He rejects the Aristotelian concept of nature and replaces it with that of the soul. If one still wants to employ the term "nature," one might say that nature is the impression which the soul makes on matter. The soul is non-spatial, it is not within things, it is nowhere. For this reason it needs heat as an implement for its workings within things. Heat itself is an active principle; it is the cause of all motion. But by themselves its workings are purposeless and destructive; it destroys the universal order. This explains the presence of heat, for example, in the processes of decay and decomposition. All beginnings in the universe are the result of the interactions of matter and moisture and of soul and heat. Should either moisture or heat be absent, or should heat not be directed by the soul, destruction ensues.

Hence, all permanent bodies, including stones, are always slightly moist and warm and of necessity animate. But here the soul is in a way only partially conscious and moves toward its goal in an almost dreamlike manner. Life and substances associated with it can interfere with the workings of the soul. Thus, the soul might well err in its explorations and choices—there is no doubt about this in Cardano's mind. This accounts, for example, for the genesis of various kinds of monsters. But because the soul exists outside of space and time, it is itself not disturbed by this. Altered are only its effects, much as the rays of the sun are eclipsed by the moon or by clouds without the sun being adversely affected.

The goal of all formative activities of the soul is the human form. However, it moves toward this goal gradually and tentatively. Thus, it has produced animals which are more or less anthropomorphous: apes, cats, dogs. All are structured according to a single plan. For example, they all have paws with five digits. The more man-like these animals are, the more useful these digits become. In the case of dogs, the five toes serve no apparent purpose. They have meaning only as a preliminary form of the human hand. That the soul works with only a general notion of the goal of its activities is further evidenced by the fact that men, like women, have breasts, although they have no need for them. The reproductive organs of both sexes are initially positioned inside the body; in the case of the male they are, because of the intense body heat, later pushed to the outside. "Everything is planned analogously," Cardano says, "but not everything is realized analogously." In other words, the soul only clarifies its goals in the course of its activities. Since everything is animate, Cardano has to explain the difference between stones, plants, animals, and human beings. These differences are mainly due to the fact that heat, the working tool of the soul, can be present in matter in varying degrees.

Life can only be generated by very intense celestial heat. Although stones as well as plants are in some measure sentient, only animals hear, remember, and in some subconscious way also have mental images. In man, however, the soul is linked with the rational spirit or mind ("mens"); thus a being evolves that perceives scent and taste, light and darkness, pleasure and pain. It has memory as well as imagination, it sleeps and wakes, and it never returns to its previous state. Only man has self-awareness while at the same time realizing that others are aware of him too. Animals do not know themselves and do not understand what they do.

Next, the question is raised how a single soul can be an adequate principle of such a multiplicity of forms. This question is all the more relevant since Cardano's theory of the soul is decidedly Averroistic and contradicts the Christian doctrine that every human being has been given an immortal soul. Cardano extricates himself from this dilemma with an argument which we later encounter in much the same form in Leibniz.

Cardano takes as his starting point the conception of the soul as something infinite, comparable with an infinite number in which all numbers are potentially present. But whereas in mathematics individual finite numbers actually exist, while the infinite number exists only potentially, the reverse applies to the ulterior soul. Here, the infinite is real. One might compare the soul with a continuum in which the parts are potentially present. One must keep in mind that the soul is complete actuality; nothing is contained in it potentially. This implies that the parts of the soul are also actual. The soul is an actual infinite unity and at the same time an actual plurality. The individual souls are not merely potentially present in it. The unity of the soul is realized in its order, which requires plurality.

From this Cardano, like Leibniz, deduced a theory of freedom of action and of chance.[5] Cardano neither doubts man's freedom of action nor the occurrence of random events. In human affairs chance seems to be an important ally of success.

Yet it must be noted that the same event might well be seen as fortuitous by one person and as predictable by another. There are many gradations, as there are between hot and cold. Hence there must exist a highest, perfect insight which is independent of good fortune or chance; such insight is divine. Chance and good fortune are based on the law of causality; the knowledge of this order is divine; therefore, chance and good fortune seem to us to be of divine ordinance.

This being the case, little seems fortuitous to the philosopher. But no man possesses comprehensive knowledge; human nature is incompleteness. Cardano's explanation is essentially based on the idea that the soul and the order which is established by it is infinite. It is conceivable that a hierarchy of things, an order exists within the finite world which man can perceive.

In this case, things appear to be correlated in accordance with certain laws. On the other hand, such correlations may be only discernable through an analysis of such vast scope that no individual is ever likely to be able to carry it out. Ultimately, unity of all things becomes apparent only in the infinite. Such insight is completely beyond man's grasp. It can be attained only by God as He is Himself infinite. To man, such continuity appears to be governed by divine ordinance.

I have tried to present Cardano's philosophical ideas and reflections in as orderly a manner as is possible. It has become evident that his ideas are based on a logical system which rivals that of Aristotle. His conclusions are by no means arbitrary, but follow logically from his assumptions. These supplement each other in a rational way. Thus, the overall picture is rationally coherent as well as imaginative. I have frequently used Cardano's own terminology, since it is generally descriptive and exact. But I have not followed him at all in the organization of my presentation because he basically wrote without any disposition.

I would, however, like to give the reader just an idea of the actual style of *De Natura*, and am therefore transmitting the conclusion of this treatise in as literal a translation as possible.

Cardano says:

We state, therefore, that the incorporeal is infinite and one, like a light, and its cause is God. The corporeal, on the other hand, is finite since it can be measured in terms of length, breadth and depth. Light, as it streams forth, is a continuous unity and it is separate from us. There is, then, in the incorporeal neither division nor number; it is just the One, although it can manifest itself in gradations of perfection. Every material substance, however, has a boundary. Experiences such as running, carrying a weight, seeing, are all finite. There are simply physical phenomena, or phenomena which cannot occur outside of matter—like memory, for example. But thought and will are infinite.

Now, if matter is governed by the infinite, and if the immaterial streams forth from the incorporeal source, then matter also emanates from the incorporeal. Inasmuch as a material substance is such an emanation, it is also infinite. Yet, it is finite when thought of in terms of length and breadth and depth. Therefore, it is part of something, just as the heavens in their perpetual motion are themselves a part: matter is itself contained in something else, contained in a space, as a center of the incorporeal. As the incorporeal moves away from its origin, it degenerates. As it intermingles with matter it becomes soul: heat mixes with the opaque, but light remains separate. Most divine is what never intermingles with anything else—the mind, or rational spirit, for example. That which intermingles with the soul is eternal and incorporeal. The incorporeal cannot be mixed with any material substance since it neither possesses any quality nor can it be divided. It follows that no soul can be contained within a material substance.

What, then, are the differences between the soul and the mind? While the mind does not react with matter or qualities, the soul is connected with life as the mind is connected with the soul. The efficacy of this connection determines the gradational perfection of the workings of the soul. It will be different in the act of procreation, in a tool, in the excellence of a physical shape, be it that of a lynx or that of a fish. It will differ with regard to the faculties of vision and hearing of the lynx, it will be different with different persons—for instance, with someone who has not eaten as opposed to someone who has eaten well. It is apparent, then, that the souls are uniform with regard to their origin but not with regard to their perfection. Closer to the body the shadow is denser, getting less substantial toward its outer edges. The order of such differences corresponds to the order of the material substances.

Every human being can be governed by reason since the mind transcends life. Yet the mind cannot move a soul immersed in life, since such a soul is too heavy.

Animals do not have the power of reason because they do not have rational spirits, that is to say, they have nothing that could move their souls. Like water in a container, the soul is moved by life.

But why is the soul connected with life in man? Because the order is imperfect! But, taking distance[6] into account, there can be no order more perfect than the one that connects life and the human soul. The infinite is indeed totally incorporeal, just as man is incorporeal when he is not bound by his senses. But even the infinite has boundaries when one thinks of it in terms of its parts. Thus, the formation of the organic body constrains the psychic substance like a harness. The perfections of the substance correspond to those of its source since it is governed by this. One might say that the soul generates life and carries out the workings of nature appropriate to the material, provided its instrument—solar heat—is present. After all, solar heat and physical substance extend simultaneously. In accordance with the degree of perfection, the body will by its own activity affect that which complies with the demands of the soul.

These final observations stress once again the fundamental difference between mind and matter. The latter is essentially finite and limited, the former is actually boundless. In taking this position Cardano departs altogether from Aristotle, for whom the infinite souls exist only potentially, and follows Platonic doctrine.

The above statements are followed by some rather peculiar reflections upon the emanation of the corporeal from the incorporeal. The material thus produced is infinite, with the cosmos as its center. Cardano seems to entertain here the notion of an infinite space—partly spiritual, partly physical—in which the finite cosmos rotates.

Related to this theory of emanation is the idea that the soul degenerates as it proceeds from the One, or to put it differently, that the soul is produced by a degeneration of the spiritual as the latter mingles with the material. This neo-Platonic concept does, however, contradict all of Cardano's other assertions. Further on he reasserts that no soul exists within the body. Yet, the mind is something higher or purer than the soul. This means that the soul must in some way become adulterated. Therefore, Cardano states that the soul is connected with life as the mind is connected with the soul. Although this is certainly an imperfection, it is an unavoidable one if a connection between the spiritual and the material is to be established at all. One might say that at this point Cardano's theory becomes flawed. But no philosopher has yet been able to surmount this obstacle.

Mathematical Theosophy

Cardano's philosophy of nature may be regarded as an original if somewhat eccentric variant of Renaissance Aristotelianism as it flourished in Upper Italy, particularly in Padua.[7] But there are also indications that in later years Cardano became familiar with the writings of Nicholas of Cusa, and he seems to have been rather impressed by this remarkable figure. I did not find any direct references to the philosophy of the cardinal, only to his mathematical writings. In his inaugural address, "laus geometriae,"[8] which he held in Milan in 1535, Cardano referred to Nicholas of Cusa as a most ingenious man who did, however, almost always arrive at the wrong conclusion. In book XVI of *De Subtilitate*, Cardano praises the value of algebra, pointing out that it enabled Monteregius to disprove the calculations made by Nicholas of Cusa on the circumference of the circle. Cardano obviously knew Monteregius's *De Triangulis*, which includes the faulty calculation in its appendix. The book was published in Nuremberg in 1533, making it quite up-to-date at the time. Cardano must subsequently also have come across the philosophical writings of Nicholas of Cusa, perhaps in the edition published by Henri Petri, Cardano's own publisher, in Basel in 1565. These writings apparently made quite an impression on Cardano. In 1570, the *Liber de Proportionibus* was published. It concludes with a quasi-mathematical theosophy, containing various propositions which Cardano partly proves and partly refutes. These propositions seem to have been directly suggested by the ideas of Nicholas of Cusa; they are certainly not part of any traditional school of thought. But Cardano does not mention the cardinal by name, perhaps he did not want to contradict the eminent personage openly.

Cardano's theology is interesting in itself and characteristic of the time, but it is also rather interesting to trace the influence which Nicholas of Cusa had on the ideas of an eminent mathematician. One is reminded here that Newton concluded the second edition of his *Principia* (1713) with the famous "Scholium generale," which deals with theology rather than physics.

First, I would like to describe the *Liber de Proportionibus*,[9] because it seems useful to know the framework in which this curious mathematical theology appears. The *Liber de Proportionibus* is a mathematical work of two hundred seventy pages in folio. Cardano designated it as volume V of an encyclopedia of mathematics—the *Ars magna* being volume X—of which he wrote substantial parts; however, only little of this work is still extant.[10] As is customary in mathematical writings, the theses are put forth in the form of "propositions" (there are 233), followed by a "commentary" which is supposed to furnish the proof. Cardano begins with purely geometric and arithmetic statements of the theory of proportion. These are followed by applications to questions of physics as well as to problems in medicine. Many examples deal with mechanical questions, some theoretical, others of a more practical nature. Not all of the assertions made are correct. Cardano stays basically within the Aristotelian-Scholastic tradition, which Galileo was to break with two generations later. The book is discussed in works on the history of mechanics as a characteristic example of this period of transition. But Cardano himself does not intend it as a treatise on mechanics; instead, the work is supposed to demonstrate that all of reality can be basically represented and explained in mathematical terms. He takes his examples from many different areas, and although the sequence of the propositions is not entirely systematic, the discourse moves on the whole progressively from earth to the higher spheres, ending with the divine. This highest sphere, however, cannot be expressed in conventional mathematical terms. It requires a special, symbolic mathematics. On this point, Cardano agrees with Nicholas of Cusa, but he differs with him in his conception of symbolic mathematics.

I would like to discuss proposition 183 as an example of the way in which Cardano applies mathematics to "natural" problems. This proposition does not pertain to physics but rather to medicine. Proposition 183: "The natural span of life is to be explained by its fortuitous duration."

As is often the case with Cardano, the meaning of this proposition becomes clear from its solution. He assumes that man has an innate lifespan which is also the reserve of the vital energy used in the course of a lifetime. It is expended first of all by time, so that with each chronological year a year of the reserve is consumed. But people draw on their reserves in addition to that, especially in their youth. They eat and drink too much, they worry about trifles, they work too hard, and they yield to every conceivable passion. One can assume that even a moderate and sensible person will thereby use up an additional one-fortieth of the reserve he has. Experience shows that such persons will live close to eighty years unless an accident or a severe illness causes untimely death. Based on this maximum lifespan and using a method similar to the computation of annuity,[11] Cardano calculates the life-reserve for each year of a person's age. At birth, this reserve is 260 years. The figures for each year are given in a chart. It shows that at the age of sixty a person still has a reserve of twenty-five and one-half years. Even if he decided at this time to take up an extremely moderate and dispassionate way of life, he could no longer extend its duration to any considerable degree.

Cardano believed that his theory was actually rather well supported by experience. He disputed the objection that the initial reserve of 260 years is so great that one ought occasionally to meet persons who have reached the age of 200, by arguing that such a person would have had to lead an extraordinarily moderate and dispassionate life from his earliest years on and yet grow up to be healthy and vigorous. Cardano did not believe this to be possible.

It should also be noted that a sensible and moderate way of life does not mean that one must renounce all pleasures. It is the excesses which are harmful, particularly during one's youth. Furthermore, moderation is by no means a guarantee for a long life, since a person can meet with misfortune at any moment. Nothing in life can be taken for granted. But one need only think of the deep insight into the ways of the world one can gain in the course of a long life to realize that it is well worth the effort to free oneself from excesses and passions. Cardano's theory quite obviously serves medical purposes which we would today call preventive medicine. Although it is not a theory in the modern sense, it is certainly interesting and also quite sensible if one considers its pedagogical aim.

Proposition 221, which deals with the size of the moon and the stars, concludes the part of the book concerned with the natural sciences. The subsequent fifteen pages concern the immaterial world, and in my opinion they are also an analysis of the ideas of Nicholas of Cusa.

A book on proportions would indeed be an appropriate place for such a discussion, since Nicholas of Cusa states at the very beginning of his *Docta ignorantia* (Book I, chapter 1) that all research is guided by the idea of proportion. In accordance with this view he attempts to express the divine by a kind of symbolic mathematics. In its context, a special role is assigned to the infinite. In this realm of the infinite the principle of contradiction no longer applies: God is the "coincidentia oppositorum."

Cardano shares with Nicholas of Cusa the belief that the divine can be expressed in the language of symbolic mathematics, but he differs with him as to the way in which this ought to be attempted. Cardano prefaces his views with three definitions and four axioms in order to characterize symbolic mathematics as he understands it. He does not disregard the principle of contradiction, but he rejects the idea of the infinite as actuality; he also changes the concept of proportion. He states that common or actual proportions refer to quantities. Immaterial substances, however, have no quantity. Therefore, he gives the definition: "A sublime proportion or order is the comparison of substances which have no quantity."

This seems to imply that sublime proportion also produces order like the ratio of numbers, but that this is not a numerical order and that it has nothing to do with concepts of "greater than" and "less than."

The axioms read as follows:

1. The infinite, although it has the semblance of quantity, neither has nor is quantity.
2. The contradictory is that above which there is no power. (Repugnans est super quod nulla est potentia.)
3. That there is no power above the contradictory does not signify imperfection, nor does it repudiate the notion that the infinite does not exist. (Non posse super ea quae repugnant, nullam declarat imperfectionem, neque infinitum non esse negat.)
4. One infinity cannot be greater than another.

The second and third axioms refer to the "repugnans," that which contradicts logic. Five cannot possibly be even, since the concepts "five" and "even" are contradictory. Even God Almighty cannot change that. But this does not reflect upon God's omnipotence. The above axioms assure the full legitimacy of indirect proof,[12] as it is based on the demonstration that the negation of an assertion leads to a contradiction, thereby proving that the assertion is correct. In the subsequent development of his ideas about the immaterial world Cardano makes frequent use of the procedure of indirect proof.

The first and fourth axioms concern the infinite. It is certainly not infinite in terms of quantity, and—according to the third axiom—no indirect proof of its existence can be given either; this is emphasized by the triple negative neque-non-negat. In my opinion these axioms are clearly intended to refute the views of Nicholas of Cusa.

Armed with his axioms, Cardano now attempts to explain the incorporeal world. It contains time and eternity as well as the "vitae" (vital principles or souls), the "mentes" (rational spirits or intelligences), and finally God or the divine personages Father, Son and Holy Ghost.

The "vitae" are not only the souls of humans and animals, but also the souls of the celestial spheres which Aristotle calls the "movers," a form of angels. The "mentes" are a higher form of angels, as they no longer have any direct connection with the physical world. The concept of "aevum" (eternity) is explained in *De Subtilitate.*[13]

On time and eternity Cardano remarks in

Proposition 224: "Neither all of time which we conceive of as infinite, nor the aevum vitarum is in any way proportionate to a potential time period such as a day or a month." The proof simply states that time is either actuality or relative to actuality, and that its parts are potential. If the whole of time were not actual, then nothing would be actual, which is absurd. Therefore, time is actual. The parts of time, however, are merely potential. No proportion can exist between that which is actual and that which is merely potential. Cardano is cautious, though, merely stating that we conceive of time as infinite, not that it is infinite. Its actuality is proven indirectly.

Proposition 225: The revolution of the celestial spheres takes place within time. This raises the question of whether the speed and size of these spheres might not allow us to make a quantitative deduction with regard to their movers, the "vitae." As is to be expected, the answer is negative, since the movers do not move by strength but by will. This will is directed by the goal which God has set for the world. Consequently, all these forces are working toward one goal, and the extent of their apparent workings permits no quantitative deduction concerning themselves.

Proposition 226: Yet there exists among them a sublime proportion. This is determined by their proximity to or distance from the First Cause, that is from God. The incorporeal is neither finite nor infinite, neither extended nor "contract," since these are all references to quantities. The incorporeal is devoid of quantity.

I did not translate the word "contract" because Nicholas of Cusa uses it like a concept. In the *Docta ignorantia* (book I, chapter 2), he says that the universe, which is plurality, may be conceived of as the "contractio" of the first, supreme, and absolute Being. Cardano apparently could accept neither the cardinal's conception of God as the absolute maximum which is also unity, nor his idea of the multiplicity of the world being the "contractio" of God. Such concepts are symbolic gradations of quantitative notions which cannot be taken into account here at all. One can, however, say: "The vitae have a certain proportion corresponding to the level of their perfection. More specifically: God is the primary being by virtue of his nature (per se primo); he is an absolute, the source of all good; he is existence. From the 'first good' proceeds wisdom; it is not the cause of all that is good, because it cannot produce the 'first good' and at the same time proceed from it. But it is nonetheless primary itself, an absolute and a good. Love, on the other hand, is the cause of all subsequent good; it is an absolute, it exists by itself, but it is not primary. Finally, there is life (the vital principle), which governs the physical world; it neither is an absolute nor is it primary by itself. (It is merely the source of all good.)[14] Yet it is absolute in the order of things, and it maintains this order. Thus, we speak of the deity as being several personages, spirits, and incorporeal substances."

Deus, Sapientia and Amor mean Father, Son and Holy Ghost. Sapientia has always been identified with the Son, who is designated as logos in the prologue to the Gospel according to St. John. A further frequently-quoted reference to this identity are verses twenty-two to thirty-one of the eighth chapter of the Proverbs of Solomon, where Wisdom says:

> The Lord possessed me in the beginning of his way before his works of old.
>
> . . .
>
> When he prepared the heavens, I was there.
>
> . . .
>
> Then I was by him, as one brought up with him: and I was daily *his* delight. . . .

Love is identified with the Holy Ghost, as Father and Son are united in love. And because the Holy Ghost proceeds from the Father as well as from the Son (qui ex Patre Filioque procedit [cf. the Credo]), love is "per se, sed non primo."

To these three hypostases Cardano adds life as the fourth. It is less divine: "non absoluta, neque per se primo." Nonetheless, it is absolute in the order of things, and it maintains this order and governs the world. In this context, life apparently signifies the World-Soul, which for Cardano embraces all spiritual forces. Life, as Cardano uses the concept here, also stands for the other intelligences and incorporeal substances he mentions.

At this point, the development of the argument has reached a certain preliminary conclusiveness, and Cardano takes the opportunity to extend the discussion to those sciences which are concerned with predictions and presages, some of which are based on natural, others on supernatural phenomena. He was most interested in such questions, being himself an eminent astrologer, physiognomist, and interpreter of dreams. This discussion is, however, merely a digression. Cardano soon returns to his main topic. He states in:

Proposition 229

All incorporeal things are one; they cannot be conceived of in terms of number. This appears at first to be a paradox, but it really is not, because the statement can be proven. There is no number that could express the incorporeal, as it would, of course, have to be either finite or infinite. It could not be an infinite number, because that would not denote any measure at all. This would mean that there was neither God nor any primordial substance. How could God be the master of an infinite multiplicity? How could he be the First if there was no last? But there could be no definite number of incorporeal things either. What should it be? It could not be limited by one hundred, nor by one million, nor could it be measured by a continuum. It follows, then, that everything incorporeal is one in the sense that it flows from its primary source in gradations of perfection. At its boundaries, the human mind (mentes nostrae) is joined with the intelligences and the heavenly realm. The incorporeal imparts itself to the lower regions and the corporeal substances (these are the celestial spheres) and sets them into motion. Because everything is connected with the primary source, everything is also moved by it. God could not be the all-encompassing Being (ens commune) in whom all of reality is fundamentally contained, if He were not one. He would then have properties, and to think this is wrong, absurd, even sinful. But man is a sensual being. It is, therefore, permissible for him to devise numbers in analogy to the physical in order to express metaphysical things. The physical world is an image of the higher world. It consists potentially of an infinite number of parts. Consequently, we can also conceive of an order of the higher world that consists of an infinite number of parts.

In his description of the divine personages, Cardano avoided using such words as eternal, boundless, almighty and similar traditional predicates of the infinite. The infinite is not actual existence, as is God. On the contrary, its existence is very uncertain and—according to the third axiom—cannot be proven. For this very reason God cannot be master of an infinite multiplicity, because then He would also be the master of something dubious. But since we are sensual beings, we devise numbers, and we are permitted to think of the higher world as infinite.

In his treatise *De Uno*, Cardano says: "Number is a fiction of the human mind, and its principle is the One."[15]

Some of the statements made by Nicholas of Cusa are very similar. In his *Trialogus de possest*[16] he states that God is Trinity without number, and that the mathematical number is a fiction of the human mind with the One as its principle.

Nicholas of Cusa defines numbers as "the unfolding of the intellect." In devising numbers the intellect is creative. It may, therefore, conceive of God's creation of the world in terms analogous to the creation of numbers. This train of thought was developed in *De Coniecturis.*[17] In chapter fourteen, the three worlds are described which are created as the divine unity descends. In chapter fifteen, the author says: "I furthermore surmise that each of the three worlds repeats the number series so that each is perfect in its way, although with differences of degree. Thus, from the original unity within the universe new graduated unities develop. This is illustrated in the figure of the universe below. It shows three worlds, nine chains

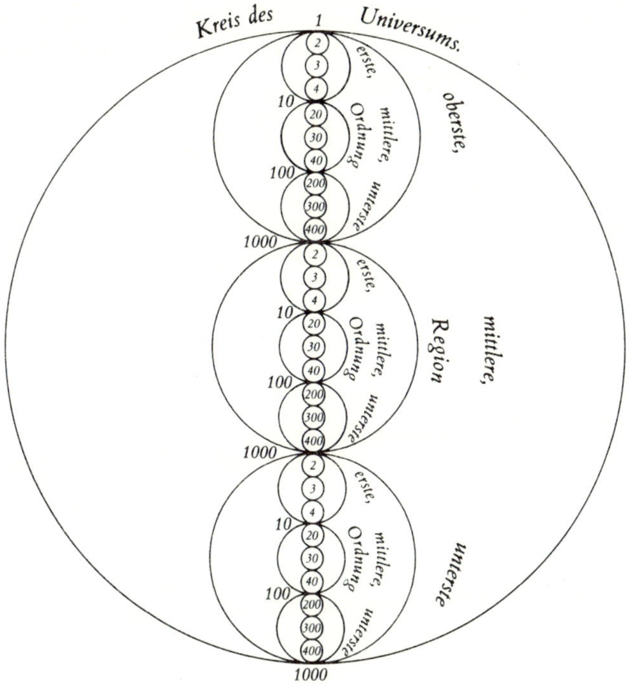

of order, and twenty-seven choruses. It is essential for the understanding of the entire theory." In my opinion the following proposition, 231, is a direct reference to Nicholas of Cusa.

Proposition 231: "There are three worlds, but no proportion exists among them. Neither can these worlds be comprehended in numerical terms." The commentary explains that the three worlds are God, the physical world, and—acting as intermediary between these two—the souls (vitarum et hominum et coelestium). However, these worlds are not three in a numerical sense; here, the concept of three expresses a notion similar to that of perfect, more perfect, most perfect, or to that of beginning, middle and end. The proportion that exists among them is purely sublime.

Cardano obviously agrees with Nicholas of Cusa on the existence of the three worlds. On the other hand, he completely rejects the attempt to comprehend them in cabalistic fashion by some form of number-mysticism, even if only by conjecture or by supposition. Instead, he calls the proportion which exists among these worlds "sublime." But what is actually meant by this is never adequately explained. The comparison with substances without quantity as a definition of this proportion is obscure. Proposition 226 states that the sublime proportion among the vitae, the movers of the celestial spheres, depends upon their proximity to or distance from the First Cause. The idea of the spheres and their movers originates in Book XII of Aristotle's *Metaphysics.* According to Aristotelian doctrine, the motive underlying the motion of the spheres is teleological, directed toward a goal. Cardano expresses the same idea. Aristotle adds the notion that this purposive cause moves things like the beloved moves the lover. A sublime, grand sentiment—rarely encountered in his other writings—permeates Aristotle's reflections on this topic. This leads me to interpret Cardano's definition of sublime in the sense of a comparison based on emotional values. Distance from or proximity to the First Cause are degrees and intensifications of such an emotional value. To the extent that these emotional values are experienced and evaluated, emotion is also a rational function. This means that the principle of contradiction applies, and that we can interpret the phrase "repugnans, super quod nulla est potentia" to mean not only that which is logically contradictory, but also that which is contrary to the emotions.

I think that such an interpretation of sublime proportion is consistent with Cardano's concluding remarks.

Preceding these final thoughts is Proposition 232 with an extensive and rather muddled commentary.

Proposition 232: "The greater its speed, the more each natural motion resembles a state of rest. This assertion appears at first to be false. It is therefore important to understand the exact meaning of the words in this context. That the assertion is indeed correct, is supported by authority, by perception, and by twofold logical arguments." A lengthy and often curious discussion follows. One of the examples used to support the proposition is the following: "An object rotating at infinite speed would, in terms of its parts, always remain in the same position, that is to say be motionless." Cardano explains: "If one were to look at the wheel used for sharpening swords, one could observe its motion, provided it was not too fast. However, as the acceleration reaches the point where the motion is no longer discernible by the senses—and provided the wheel does not diverge from its axis and thus starts to shake—it will give the impression of standing still. Such a perceptual judgment would be further reinforced if the speed were to increase to the point where one could no longer distinguish between beginning and end."

Some further discussion leads to the conclusion that in the realm of the eternal, motion is equivalent to rest. Consequently, the incorporeal ought to be thought of as changeable whenever it is—like light—joined with the corporeal.

The major part of the discussion consists of peculiar reflections on kinematics which are not really relevant to the topic we are concerned with. But the proposition and its supporting arguments are relevant to our discussion, because the assertion made by Cardano about the circle rotating with infinite speed and being at rest is made in almost identical form by Nicholas of Cusa in his *Trialogus de possest*.[18] There, the question to be answered is how the eternal is everything at once while the whole is contained in the "now" of eternity. To explain this, the cardinal uses the example of a child's spinning top, seeming to move less the faster it turns. At its maximum speed it would be at rest. If one thinks of the circumference of the spinning top as a circle, at infinite speed each point of the circumference would be everywhere simultaneously. Therefore the cardinal states: "The greatest motion is also the least and none."

These reflections are particularly characteristic of the symbolic thinking of Nicholas of Cusa. The image of a circle at rest while moving at infinite speed is uniquely his.

Cardano's proposition basically restates the one made by the cardinal. When Cardano says that his assertion is "authorized" (haec propositio . . . autoritate . . . manifesta est), Nicholas of Cusa is the only authority he could be referring to.

Despite certain reservations, the final propositions evince Cardano's respect for the cardinal. He gives the following summary of his thoughts:

> All that exists in the incorporeal world is eternal, blessed and certain. It is immutable only with regard to its place; its essence is becoming; it is caressed, as it were, by the soft murmur of water and the summer's breeze.
>
> All that exists in that world is neither part nor whole, because it would then be a quantum, a definite number. It knows no change in time or space since it is not contained within them. It can, therefore, neither acquire nor lose anything. It has neither beginning nor end. As it proceeds from

the First, it is highest delight, comparable to the ecstacy of those who have achieved true insight and happiness. For the very reason that such ecstacy is perpetual and certain, it is not without variety; this makes it still more glorious, just as earthly pleasure is made more attractive by the variety of its causes and effects. Thus, in a continuous process of giving and receiving, it is always novel; it is one, and it is always actuality. What exists beyond that world is the potential, the basis of the infinite as imagined by the soul. Since this is not ordered by the primary order preceding all beginning, there can be no doubt that the infinite is not a cause. The soul, entangled in matter, strives for order. Whenever the soul finds such an order it rejoices, because in such an order the soul recognizes the inner purpose of things. This is evidenced, for example, by the diverse properties of numbers. The concept of potentiality underlies our idea of the infinite because we imagine that there could be an infinite series of phenomena. This means that potentiality is to be thought of as an imcomplete process. As our soul moves away from God, it faces things behind it. They harbor potential imperfection, disorder, danger, the infinite, even despair. Everyone who turns away from God will experience this: the more he acquires, the more he will feel is lacking. As the number of sons increases, as wealth and reputation grow, fears and troubles will multiply. This element of novelty [in the incorporeal world which Cardano means when he says "thereby it is always novel"] has been interpreted by some as a ceremonial dance, by others as music of the spheres.

What has become apparent, then, is this: the substance of the incorporeal world is engaged in a constant change of order; it is without motion, time, or place.[19] From this springs mutual love, and delight, and the totality of the One. Thus the soul, having recognized God and the heavens, descends and creates new order. There is such great bliss in that other world that we cannot compare it with ours which can be nothing but its shadow, even if it were pure, even if it were permanent.

Therefore, let this be the end of our reflections on the nature of the divine—and the end of our book.

4 *De Subtilitate* and *De Rerum Varietate*

DE SUBTILITATE is considered Cardano's chief work, although he himself hardly shared this view. Certainly his natural philosophy cannot be inferred with sufficient clarity from this work alone. For this reason Cardano suggested in *De Libris propriis* that *De Natura* be studied in preparation for this book. It then becomes apparent that the essential ideas of his natural philosophy are already contained in *De Subtilitate,* albeit frequently without any logical explication. Furthermore, his anti-Aristotelian position is not nearly as apparent in this work. One must keep in mind that the educated reader of the time was well-versed in Peripatetic philosophy, and was—even without explicit references—probably much more aware than we are that Cardano did not always share the opinions of "the philosopher."

The book *De Rerum Varietate,* which was published in Basel in 1557, can be viewed as a supplement to *De Subtilitate* in much the same way that the second volume of Schopenhauer's *Die Welt als Wille und Vorstellung* is a companion volume to the first. *De Rerum Varietate* contains materials which Cardano gathered in preparation of *De Subtilitate,* as well as supplementary materials which he considered important. It probably did not seem practical to combine all this material with that presented in *De Subtilitate,* since the latter work comprises in the improved and enlarged Basel edition of 1560 an impressive 1430 octavo pages. The first edition of *De Rerum Varietate* is an equally imposing folio volume of over seven hundred pages. Together, these two works are quite extensive and comprehensive. These are not theoretical works. Rather, they are meant to unfold the richness of the world—as well as Cardano's profuse knowledge—before the reader.

To try to sum up the content of such books would be a futile undertaking. Still, I shall try to convey to the reader at least a general idea of their content and style. As I already mentioned, *De Subtilitate* in particular enjoyed enormous success in Cardano's time and continued to be frequently reprinted—and that means widely read—throughout the seventeenth century, long after Cardano's death. With this book Cardano exerted a marked (although rarely acknowledged) influence on the intellectual life of the baroque period. *De Subtilitate* is an encyclopedia "of all knowledge," organized according to the same principle used in the arrangement of the writings of Aristotle. The work is divided into twenty-one books, each consisting of between twenty and thirty double-columned pages in folio in the *Opera* edition. The titles of these twenty-one books offer an initial overview. They read as follows:

1. De Principiis, materia, forma, vacuo, corporum repugnantia, motu naturali et loco
2. De Elementis et eorum motibus et actionibus
3. De Coelo
4. De Luce et Lumine

5. De Mistione et mistis imperfecte seu metallicis
6. De Metallis
7. De Lapidibus
8. De Plantis
9. De Animalibus quae ex putridine generantur
10. De Perfectis Animalibus
11. De Hominis necessitate et forma
12. De Hominis natura et temperamento
13. De Sensibus ac voluptate
14. De Anima et intellectu
15. De Inutilibus subtilitatibus
16. De Scientiis
17. De Artibus artificiosisque rebus
18. De Mirabilibus et modo repraesentandi res varias praeter fidem
19. De Daemonibus
20. De Primis Substantiis seu vitis
21. De Deo et Universo

As can be seen, the first twelve books deal with the material world, beginning with the principles and elements, proceeding to inanimate objects, then to animals, and concluding this series with man. The subsequent nine books describe the immaterial world, again in ascending order, starting with the senses, proceeding to the activities of the intellect (a discussion of the "useless subtleties" precedes that of the sciences), and concluding with the spiritual powers, demons, angels, and finally God Himself.

In an introduction to the first book, Cardano remarks on the difficulties of the task he has set himself. Obviously, the introduction—and indeed the very title of the book—were meant to generate publicity; they were intended to rouse the reader's curiosity and make him buy and study the book. The work did indeed find many readers, and for many years, was an indication that Cardano did not fall short of the expectations he had raised. I will now present the beginning of this introduction as a characteristic example of Cardano's style.

The task we have set ourselves in this work is a treatise on subtlety. One might say that subtlety is responsible for the difficulty the senses have to grasp the physical and the mind has to comprehend the spiritual. If the question of the exact nature of subtlety already involves painstaking and most complicated investigations, what then must be expected of a treatise which proposes to explain the many different forms of subtlety? It will be apparent to everyone that we are dealing with the most complex issues in each discipline. Moreover, the proper presentation of the issues involves even more work than the actual research. An author is faced with various difficulties: obscurity of the subject matter, misgivings about unresolved issues, the search for causes and their proper explanation. In this book, all these problems arise in abundance. If obscurity causes difficulties, let me say that this book chooses to deal with the most obscure things. Keeping in mind that it has always been difficult to obtain certain knowledge, what could be more difficult than what we propose to do? There are some authors of whom I must beware, like Pliny[1] and Albertus,[2] who quite obviously lie and are, therefore, not to be trusted. There is really no one to emulate. If I have failed to carry out the most thorough investigations, I have—as the saying goes—wasted effort as well as breath.

What shall I say about the causes? They are still unexplored. They are conveyed to us through an oracle, so to speak, and as such should be accepted and believed without being proved. But nobody believes anything without proper proof. Now, I am to explain single-handedly matters which during centuries of scientific investigation have never been touched upon.

Anything which has not yet or only recently been discovered

does not have a name, and no definitions are at hand. It is indeed most difficult to find the proper terms for such new things in a dying language.[3] If I go ahead and invent names and concepts, I am obliged to explain myself publicly to avoid being challenged by those who have written recently about the same matters. Besides, even Oedipus could not give satisfaction to the reader. Thus, having undertaken such a difficult task, I hope that, more than the great effort, the product itself shall be accorded the prize of usefulness and renown. There are other things the Ancients did not deal with properly, but I shall not address myself to these matters, for I hold that anything written which cannot be substantiated by experience is of no consequence. Such then are the difficulties which arise in writing a treatise of this kind. But to get back to the subject at hand, many obscure things are indeed of a most delicate and subtle nature, but they are so not entirely and not at all times. Some things appear to be mysterious and complicated merely because of the way in which they are described, not because of their nature—like those knots, for example, which by chance run back into themselves, or all those things which appear extraordinary to the senses, although they are really quite ordinary. Such things do not deserve to be called subtle. Into this category would also fall the slender legs of certain people if they are due to an early undernourishment or some other extraneous detriment.

The introduction begins with a kind of definition of subtlety, and then proceeds to describe the difficulties of the task. Since "all forms of subtlety" are to be explained, there are no models. There are, of course, the great encyclopedic works of antiquity and of the Middle Ages, the works of Pliny and Albertus Magnus, but these authors are unreliable and one must not follow them. Furthermore, the problem posed by Cardano is entirely novel and actually demands a new language. But Cardano has to write in Latin, which is somewhat like filling old skins with new wine. This was bound to bring Cardano into conflict with contemporary scholars of philology, as indeed it did. The illustrious Julius Scaliger attacked him vehemently in his *Exotericae exercitationes de subtilitate adversus Cardanum,* not least because of the terminology used by Cardano. To later editions of the work Cardano appended an "Actio prima in Calumniatorem librorum de Subtilitate," wherein he defends himself, rejecting and refuting all reproaches. The reference to "other things the Ancients did not deal with properly" belongs into the earlier mention of Plinius. Cardano then returns to "the matter at hand," that is, to the explanation of the term "subtlety," which he sets apart from the way in which it is applied to things that only appear to be subtle. The example of the thin legs clearly betrays the physician.

Next, he discusses the senses and sensory illusions, then substances, that can be either self-sufficient or dependent. "There is only one single independent substance: the most supreme and unfathomable God who created the universe." The last book concerns Him, crowning Cardano's treatise, so to speak. The final question to be addressed is that of the order of the universe, of which Cardano says: "It exists in time, or time exists in it."

The Material World

And now he addresses the subject promised by the title of the first book. He discusses form and matter more or less in accordance with the Aristotelian doctrine which he was to oppose later on in *De Natura.* On the term "place" he remarks that the place of the universe is eternal, whereas material substances change their places. He goes on to say:

The place where Alexander sat at Babylon or Susa, was then in the open air, in a town, in a house, but now it could be in a field or even underground. The rostrum from which the most gifted Cicero held his orations was then above the ground, surrounded by air. But perhaps the earth has now risen and that place is now beneath the ground. And in the course of time there will be in every place countless numbers of people, as there have been already, assuming Aristotle is right in saying that the world is everlasting. . . . Wherever we direct our step we are always moving within the Eternal. Even the place in which I am now writing is eternal, and perhaps many a king or wise man has been here before.

Although Cardano treats the concept of space very much in the Aristotelian sense, he intersperses his discourse with reflections which express a new emotional value: space is eternal and establishes a communion with the kings and sages who came before us. He conveys this thought most vividly, and one can picture him as he sits and writes and lets his imagination wander. He alludes here to the famous passage from Aratos's didactic poem, "wherever we might wend our way, we are always dependent on Zeus," an expression already referred to by Paul in his sermon at Athens[4] and—in connection with Paul—subsequently cited time and again by all those for whom space was essentially divine. No less a man than Newton is to be counted among them. In the case of Cardano it is, one might say, a passing thought which he does not pursue further yet phrases quite distinctly.

The second book deals with the elements "which appear to be animated in some way." As we know, according to Cardano there are at most three elements, since fire is ignited air. All elements are essentially cold, since "all heat comes from the heavens, and therefore from the

Soul or from the light." In addition, there is moisture, which derives from the elements.

Precisely because Cardano does not consider fire an element, he discusses it at length, since its particular nature has to be described and explained. Therefore, he discusses fire, smoke, even the construction of chimneys; he talks about bituminous or sulphurous sites where fires burn which are difficult to extinguish; he explains the usefulness of the bellows, and he discusses lightning, the hottest fire, which kills all living things instantly. Man is the only exception, yet he too seldom escapes destruction. Thus, it was by a miracle that John Marie Cardano survived a stroke of lightning.

Among the other things described are fire machines, gunpowder, bombs, cannons and mines, as well as many more things having to do with fire and heat.

Finally, Cardano declares: "Decay, too, is a form of heat as well as being a productive process. First, moisture is produced, then fungi, then all kinds of plants, and finally worms and snakes. This illustrates the essence of heat which is a kind of soul."

In the following books, nature is described in a similar fashion. Here, Cardano includes a number of wondrous tales circulating at the time, which—being in everyone's mouth—were regarded as verified by "multiple experiences." Nevertheless, Cardano is discriminating. Thus, he regards as nonsense the notion that an application of garlic or the presence of a diamond will render a magnet ineffectual.

The reader might be interested to know what Cardano thought of the chemical-alchemistic theories of the time. In those days, the Arabic theory that mercury and sulphur are the basic elements was thought to be correct. (Paracelsus later added salt.) Mercury was not considered a metal because it is liquid; it was instead described as "metallic." Cardano discusses it in book V, which is entitled *De Mistione et Mistis imperfecte seu metallicis*. Book VI deals with the metals proper. Here, Cardano says:

> It is widely believed that metals are composed of sulphur and mercury because they smell like sulphur when they burn—this is especially true of copper—and also because they assume the consistency of mercury. However, two already existing substances cannot become a third. It follows that the metals are not composed of mercury and sulphur. Then there are those who believe that metals are transmutable. This has been observed in certain kinds of herbs, and it would, therefore, not seem unnatural if it also applied to metals. But this is not the case. Not everything is transmutable. Iron and copper, although similar in weight and hardness, cannot be transformed into each other. No metal can be transformed into gold or silver. There is a chance that one might succeed in changing silver into gold. Metals derive from the same matter as sulphur and mercury. But so do man and woman originate in the same blood and in the same place, yet one cannot be produced from the other. In other words, things of common origin do not later become interchangeable. Even if metals were derivatives of sulphur and mercury, they still could not be transformed one into the other. A Venetian apothecary by the name of Tarvinius allegedly changed mercury into gold, in the presence of administrative authorities and scholars, and this wondrous occurrence is still remembered. But whichever way this may have come about, it is quite certain that mercury cannot be changed into gold.

Cardano expresses his doubts about the contentions of the alchemists in no uncertain terms, and he also makes it perfectly clear that he believes the reports circulating in Venice about one metal having been changed into another to be fraudulent. And he gives reasons for his doubts. First, he gives the reasons for the initial misconception that metals are composed of mercury and sulphur; then he explains why there is no basis for the belief that one metal can be changed into another. His explanation is, of course, only convincing if one believes, as Cardano did, that stones and metals are also animate and therefore, in a sense, also organic substances. Only then is the example on which he bases his counter-argument—that man and woman, although of common origin, are not transmutable—logical and to the point. Man is the most perfect manifestation of nature; he is the final goal the world-soul subconsciously pursues in the course of its workings; therefore, man is a microcosm. One might expect Cardano to describe man in terms of such a microcosm when he discusses him in books XI and XII. Instead, he decribes him in a perfectly realistic and scientific way as an animal and social creature. Of course, man is seen as having divine characteristics as well as human and bestial ones. He is god-like insofar as he neither deceives others nor allows himself to be deceived; he is human as he deceives others but is not himself deceived; and he is animal-like when he does not deceive anyone but is himself deceived. Aside from assertions of this kind, Cardano does not at all emphasize man's status as the crowning work of creation. But then this is quite in keeping with a view of the essence of human nature as "quod decipit nec decipitur." In a similar vein, the passage "On the inevitability and form of man" first informs us that man is not a beast because he partakes of the rational spirit (mens). Then, a table of man's physical proportions is given, followed by a wide-ranging discussion of customs and religion, in which Judaism, Islam, and Christianity are juxtaposed and evaluated as to their respective advantages and disadvantages. Another question which is raised is the origin of large cities. Illustrated by various examples, Cardano furnishes a partial answer.

Finally, the discourse turns to language. Here, Cardano remarks on the happy circumstance that German and Greek have a special facility of forming compounds. Some observations on cannibals and giants conclude this discussion of "the inevitability and form of man."

Book XII, "On the nature and temperament of man," is something like a textbook on physiology, a field in which Cardano can undoubtedly speak as an expert. I am going to render a larger portion of the text in translation in order to convey the manner in which Cardano unfolds his extensive knowledge and how he is guided by a basic overall plan which does, however, often become totally obscured by his vivid imagination and the sheer delight he takes in the abundance of phenomena.

XII. *De Hominis Natura et Temperamento*

In those cases where the paternal seed dominates that of the mother, the children will resemble their father mentally and their mother in all other respects. If the paternal semen dominates the menstrual blood, they will resemble the father physically as well. If the paternal seed is prevailed over, the children will still resemble their father mentally but their mother physically.

Unlike most animals, human beings can mate and procreate at any time of the year. A single birth enables a woman to nurse continuously. I once knew a woman who after one birth nursed three brothers over a period of six years. Of course, the farther back the birth dates, the more inferior the quality of milk will be, because pregnancy and birth purify the blood.

If I should now speak of the hidden differences in human nature, I am aware that this is a broad field. There are men and women, there are the old and the young, there are different nations and different races. These groups are so different from one another that they hardly seem to belong to the same species.

The Numidians, for example, hardly ever wash their hands or faces; they don't drink water; they don't have swords; they are content with camels' milk and meat broth. They are so dirty that one might take them for animals. How crude and uncouth they appear when compared to the clean and ingenious people of Cambay in India!

Old men delight in things in which they can compete with the young, such as playing dice, for example. They don't take it lightly if they are surpassed in their erudition by younger men. Therefore, older persons enjoy the exercise of the mind, whereas the young enjoy that of the body. Their aspirations are altogether different. The old are miserly, melancholy, and timid; the young are extravagant, cheerful, and bold. It appears, then, that the interests of man at different stages of his life are as divergent as those of different species of animals. It can indeed be said that human nature mirrors the whole world, and that many and great things are concealed in its form.

The reason for this is the great power of the menstrual blood by which human nature is determined. When a woman menstruates, she will tarnish a metallic mirror, it will become rusty,[5] and she will damage the crops through which she walks. Furthermore, the firstborn son can contract leprosy if such blood gets into his bath, as I have myself witnessed. In my opinion the explanation for this is sympathy. But let us get back to the main point: I have shown elsewhere—in disagreement with the contentions of certain physicians—that all children bear some resemblance to their parents or forefathers. This might take the form of a wart, a beauty mark, body build, manner of conduct, or just the lines of the hand. Now, if these seeds blend thoroughly and if their minutest particles combine, then the children will be strong. Illegitimate children are stronger simply because the seeds are mixed more vigorously—

due to the intensity of passion. For this very reason mules live much longer than the horses and donkeys from which they are bred. And this is by no means due to refraining from mating; on the contrary, those who mate also live longer. Therefore, the reason must be the thorough mixture. Some people infer life expectancy from the condition of a person's teeth. But Augustus, who lived till the age of seventy-six, had only a few small and rotten teeth, whereas his eyes were bright and sparkling[6], as those of Alexander are said to have been. (This is according to Adamantius Sophista.)[7]

Aside from physical strength, sexual abstinence also contributes to longevity. It is part of the nature of sexual activity that it releases from the arterial blood and from the very pure spirit those elements necessary for procreation. Consequently, the body is diluted, which in turn is harmful to the brain and the nerves. It brings on the shakes, hastens the process of aging, and it also weakens the eyes. Since we mentioned the arterial blood, let us remember that we have two different kinds of arteries: those close to the surface of the skin are thin and motionless, and the blood in them is tepid and dark red. The others, more deeply inside the body, are thick and pulsating, and their blood is light red and hot because it comes from the heart. These arteries pulsate rhythmically with the heartbeat, and this motion sustains the natural body heat. It also expels all impurities which have accumulated. This heat is increased by vigorous movements, as evidenced by a faster pulse rate, shortness of breath, and perspiration.

With regard to breath: all animals breathe—some more, others less. Those who move a great deal of necessity breathe rather quickly, whereas those who move little breathe neither quickly nor deeply. Large animals such as cows, for instance, whose breathing is shallow or slow, are therefore shortlived. A reliable indicator of longevity, on the other hand, is a prolonged period of slow growth. This is the reason why elephants, humans, and camels live so long, up to one hundred years.

From my family tree, which on my father's side can be traced back 269 years, I have learned that there is hardly another family in all Italy with a similar record of longevity. The records on my mother's side support this for the last one hundred seventy years. A high degree of body warmth, a large amount of fatty body fluid, and a firm lean build are the basis of longevity. But a person with great body warmth and moisture also tends to have a bad disposition, because the warmth fosters cruelty, deviousness, inconstancy and irascibility, while the moisture makes the person sluggish, lazy, epicurean, greedy, and lascivious. Among those whose bodies are very warm and moist, intelligent people have the worst disposition, unless they devote themselves to the study of philosophy. One of the effects of diligent study is melancholy. It is caused by the decomposition of the fatty fluids due to excessive study and waking. If intelligent people will nonetheless persist in their evil and malicious ways, all one can say is that they are behaving true to their nature, and that for them the study of philosophy has been to no avail.

Dissipation of the body fluid shortens life, while its profusion lengthens life. Animals, having to move and thereby generating body heat, which in turn consumes moisture, consequently have a shorter life span than plants.

Another distinct advantage is tenuity ("tenuitas") combined with resilience. In the case of small creatures such a combination is indeed vital. Bees, for example, are of delicate build, but they are also particularly intelligent, and they live long, up to seven years. A prolonged and slow period of growth accounts for the longevity of tortoises, who can live sixty years or more. This factor does not only affect man, but all animals and even plants.

Man is a particularly delicate creature. This has been well-illustrated, for example, by the tomb of Alexander, Duke of Florence. Although constructed of very dense white marble, it nevertheless became stained because the fatty substances of the corpse penetrated the floor and dripped down onto the bases of the columns. Tenuity, then, brings with it two gifts: longevity and distinction.

Man, among all creatures, is most enslaved by the pleasures

of love, due to the great heat and moisture of his body. Only the birds surpass him in this, since they ejaculate relatively little semen, and their testicles are inside the body.

Due to their intellectual activities, intelligent men are less enslaved to Venus, because study dissipates the animal spirits and redirects them away from the heart to the brain, that is in the opposite direction of the genital organs. For this reason, these men beget weak children who bear no resemblance to them. They will greatly benefit from associating with beautiful women, reading love stories, and putting up pictures of beautiful maidens in their bedroom. They should never interrupt sexual intercourse completely, especially if they seldom engage in it.

Those who slacken easily due to looseness of body tissue will benefit from baths. Flatus, too, encourages erection. When it occurs—be it through joyfulness or compassion—the penis will become erect. I also observed in the case of men who were hanged that the penis stiffened—it was a kind of spasm. In certain people lust is aroused by the idea of inflicting pain on someone else. According to John Mirandola, this tendency can be so extreme that such individuals will achieve erection only when they beat, even whip, another person. In some people shame or fear can cause impotence. Some try to overcome this by rubbing their penis with ants that have been preserved in oil.

Cutting off the testicles will eliminate the sexual drive. This will also prevent baldness and will stimulate unusual beard growth. In addition, it will protect men from the gout. Should baldness set in for other reasons and should the hair fall out, Oleum Tartari[8] is a good remedy. It will not only stop the loss of hair but stimulate its growth. All this proves that in nature nothing occurs without due cause. Just as warmth furthers the growth of hair, so does this fine warm oil.

On account of his warm and moist body man is heavy; this is why he does not have wings. But even if he did have wings he would still not be able to fly—his arms would hinder him.

Man was supposed to have four legs, but because in his case the head is dominant, he can stand on only two legs, upright and free. Here, his particularly long feet are very helpful. No other animal has longer feet than man. But since Galen already discussed this in great detail, I need not go into it again.

This is the way Cardano describes the nature and temperament of man. Although he says that man is not an animal because he is also a spiritual being, the above descriptions quite obviously deal primarily with the animal side of human nature. Man's most remarkable attribute is his longevity, which is due to his "tenuitas." In subsequent chapters, in the course of various digressions, Cardano returns repeatedly to this subject.

As man is seen by Cardano as distinct from the animals because he is a spiritual being, chapter forty-two, "De Mente," of book VIII, "De Homine," in *De Rerum Varietate,* is an important supplement to the ideas put forth above. I am translating "mens" with "rational spirit" (mind), not an entirely satisfactory solution in view of the fact that Cardano calls the carrier of the spirit "spiritus," which I therefore render as "vital spirit." This "vital spirit" has to be thought of as a semi-material substance which is present in the human blood or in the brain. "Spiritus" is a term of Galenic physiology, whereas "mens," being a substance separate from matter, is a philosophical concept.[9]

Of it Cardano says: "The rational spirit is an eternal substance, an image of the true essence of immaterial things, that comes to man from without." The vital spirits ("spiritus") are the carriers of its workings, and if these vital spirits are destroyed, the rational spirit also loses its effect. The rational spirit is itself perpetually active. But the vital spirits tire, one reason why a person engaged in meditation often tires before having reached his goal. This goal is to restore the rational spirit to itself; this is the state of human perfection. "The rational spirit is of no specific substance, it can take any form. The ancients found a veiled expression for this with the story of Proteus." The Daimonion of Socrates is an example of how a person may experience the workings of the rational spirit.

When the mind (rational spirit) strives toward God with true passion, our human nature is raised above itself and miracles happen: The timid become brave, the unhappy joyful, the unfortunate blessed, those who were ignorant become wise, those who were weak become strong. The human mind is transported and united with the Exalted, and the body is raised with it. This passion will make one forget misfortune, hardships, and death; it will let one strive joyfully in pious devotion. But this torch is not easily lit—it is a gift from God. To open oneself for it one must first renounce possessions and all earthly desires; one must then also renounce one's relatives. One must come to despise fame as well as all material goods. One must forego the security of normal life and acknowledge one's ignorance and fallibility. One must realize that one's personal accomplishments are vain, and that all good comes from God. Therefore, one should resolve to do good, and think only of loving God and putting one's faith in Him alone. There are nine steps[10] to reach the state whereby we may become a torch of God. This Christ taught the eleven apostles and few other men.

He who has become impassioned with serving God is no longer unhappy or weak, he can work miracles, and he knows no sin. To reach this state, and even after having reached it, one must fast and pray diligently: don't eat after your labors just when the flame begins to burn, because this will impede the great fire more than water. A person whose spirit becomes associated with a bad demon will have the exact opposite experience from one who has been enlightened by the spirit. The former becomes restless, unhappy, covetous and ill; his eyes will flicker, making his state quite obvious. The divine fire, on the other hand, always produces the extraordinary.

Both conditions resemble each other, though, in that the person is roused from his usual state, the difference being that the divine fire makes men wise, pious and charitable. However, on numerous occasions this still led to an association with demons, although this was seldom noticed by the person so affected. In cases where neither of these associations—be it with the divine spirit or with demons—is fully realized, insanity will result, because the mind, once dislocated, will waver if it cannot find a new hold.

This description of the workings of the mind goes far beyond anything we find in the writings of Aristotle or his commentators, or even in Pliny. With Cardano it takes on a truly dramatic note. The impassioned vividness attests to Cardano's religious experience, and the description of the dangers that threaten man once he has been touched by the spirit are a testimony to Cardano's deep insight into the human psyche.

Thus, his image of man gains a new dimension beyond the merely innate. Cardano's assertion that "human nature encompasses the whole world," which he had so far only supported by pointing to the magic power of the menstrual blood—a rather old idea—now becomes meaningful: in man, the divine, the human, and the demonic unite.

The Spiritual World

Books XIII to XXI deal, as I said, with the spiritual world, which also includes the sciences. In book XV, before proceeding to the sciences proper, Cardano discusses the "useless subtleties." By this he means primarily magic spells and magic scripts, which he regards as ineffectual; but he nevertheless explains them in detail and with illustrations. Included here is the "Lullic Art," the use of combinatorics in the pursuit of knowledge, but he adds that closer scrutiny will reveal that there is indeed some merit to this method as it leads to combinatorial considerations. He then states that the number of possibilities of finding k out of n objects is equal to the binomial coefficient $\binom{n}{k}$, and that the sum of these coefficients is expressed by the formula

$$\sum_{k=1}^{n} \left(\frac{n}{k}\right) = 2^n - 1$$

As was then customary, these propositions are stated for an example: $n = 20$. It should be remembered in this connection that Cardano also made important fundamental contributions to probability theory (*De Ludo Aleae*).[11]

Book XVI, *De Sciencis,* opens with a series of mathematical notions and theorems. With the help of these examples Cardano tries to impart an idea of the mathematical science. In his discussion of the circle and conic sections Cardano follows Apollonius. He states that when the circumference is given, the circle will have the largest area. He mentions Ptolemy's theorem on quadrilaterals inscribed into a circle. He formulates a rule for the summation of chords in a circle which corresponds to the Archimedean theorem in *On the sphere and cylinder* I, proposition 21. He also mentions Archimedes's main theorem of a tangent to a spiral. Cardano was obviously not only familiar with the writings of Archimedes, but also understood them, something which could by no means be said of every mathematician of his time.

He then explains the "Proportio Reflexa," which he praises as his own discovery. The theorem reads as follows: In a triangle ABC, let the angle at point B be twice as large as the angle at point A, then

$$AB + BC : AC = AC : BC.$$

For this he gives a quite elementary proof.

This theorem, together with that of Ptolemy, can now be used to find the side of a regular polygon inscribed in a circle. Cardano solves this problem for the heptagon, which leads to a cubic equation. Its solution is given in the *Ars Magna.*

Finally, he discusses the then as well as later on much debated problem of the contangent angle, that is, the angle formed by an arc and its tangent. He finds that for the relation of this to a common angle, the so-called Archimedean axiom does not apply. In other words, no matter how often a common angle is bisected, it will never be smaller than any contangent angle—quite puzzling!

After this discourse on mathematics, he discusses musical intervals which can be expressed mathematically. He also talks about optics and subtle inventions in this field, such as the sundials of Vitruvius. This is followed by observations on the weather and winds, on meteorological signs in animal behavior, and on portents of epidemics. He praises the individual sciences and finally also the scholars.

Following is his list of the most eminent scholars, with Archimedes ranking first in genius and ingenuity: Archimedes, Aristotle, Euclid, John Scotus, John Suisset, Apollonius, Archytas, Mohammed son of the Arab, Heber Hispanus, Galen, Vitruvius.[12] "But all these," he continues, "are surpassed by three men who were endowed with superhuman, almost divine intellectual powers: Ptolemy, Hippocrates, Plotinus. To the last I owe the confidence in my ability to understand many things, to the second I owe the knowledge of my profession, and the first I admire because I can hardly comprehend him."

Surprisingly, no mention is made of Plato, while Plotinus is seen almost as a god-like figure. Cardano does not say, however, that he learned from him, but rather, that he owes him confidence is his own intellectual abilities. He quite obviously found his own ideas supported and clarified by the writings of Plotinus, notably by the treatise "On the soul" in the fourth *Ennead.*

Ptolemy he admired not only as an astronomer, but also as an astrologer, as the author of the *Tetrabiblos.* For Cardano, as for most of his contemporaries, astrology is a true, if mysterious, science. During the nineteenth century in particular, Cardano's belief in astrological predictions and his own endeavors in this field were viewed as evidence of his superstition. From a historical perspective, however, this judgment is unjustified. The sixteenth century was, of course, a superstitious age, and the obsession with witchcraft was rampant, especially in Germany. Toward the end of the century, the renowned lawyer Jean Bodin published his *Démonomachie* (Paris, 1580). James I, the scholarly English king to whom Kepler dedicated his work *The Harmony of the World,* also wrote a *Daemonologie* (1597), although Thorndyke[13] comments that "it is a serious piece of argument." In all of these books witchcraft plays a major role. Cardano did not share these superstitions, but he sincerely believed in omens, dreams, and the influence of the stars, and he did not really question the existence of demons. But all this still does not prove him to be truly superstitious. Phenomena are still being observed today which over the centuries have been described as the workings of demonic powers. The way in which such things are named and described very much depends on the intellectual climate and the scientific concepts of an era. An age of optimism that wants to project man as a good and sensible being is likely to use a terminology which—like a magic wand—will make even the most terrorizing experiences appear harmless by attributing them to insufficient education and lack of information.

But the sixteenth century was not yet this enlightened. Cardano's *De Subtilitate* also includes the book "De Daemonibus," in which he reports on things that were then regarded as the workings of demons. He did, however, always view these matters with scepticism. In his book *De Secretis,* in the chapter "in verbis scriptis aut figuris nulla vi esse magicam," he says:

> No proof has yet been given that demons really exist, nor that they could be controlled by means of a pact if they did exist. If demons do exist, they probably function on a higher mental plane than man and will hardly comprehend our vain ambitions and insignificant achievements. After all, humans do not understand the agreements, regulations or discords governing the life of ants. One thing is certain, though: words and images have no secret powers, they affect only the initiated in whom they evoke something. It is futile to try to discover such secret workings, although some people will experience their effect merely because they firmly believe in these things and place great hope in them.[14]

I have no doubt that Cardano believed in demons, even if he did not regard their existence as proven. The magic arts, on the other hand, which many associated with the belief in demons, Cardano viewed as self-deception.

De Subtilitate concludes with the book "De Deo et Universo." Here, Cardano says:

So far we have talked about the world and its constituent parts. It remains to discuss nature and certain hidden principles. Finally, we shall talk of God who, as the Creator of all things, rightfully deserves a distinct place and a separate treatise.

This introduction is quite unexpectedly followed by observations on climate and weather, rain and snow, hoarfrost and hail. There is even occasion to discuss the biblical manna.

Then, Cardano proceeds to explain the concepts of time and eternity ("aevum"). He says:

All things must be produced, including light, which is generated by bodies. It has itself neither body nor substance, and therefore it perishes continuously. The rational spirit, on the other hand, is an incorporeal substance and reproduces itself perpetually. All intelligent substances behave in this way, although not all to the same degree. The human intellect, for example, is not boundless. It is certain, though, that intelligences are not subject to time. They do, however, conceive of a space in which they permanently dwell: this they call eternity ("aevum"). It is analogous to the center of the sphere. The center corresponds to every point of the circumference and remains stationary as the sphere rotates. In this way, eternity remains fixed within the infinity of time. It does not expand, it does not flow, it is always at rest. God, however, is superior to all intelligences, and He exists neither in time nor in eternity. Time is never at rest. The universe, apparently at rest, is contained within eternity, and within the universe time flows.

With this, Cardano's reflections return to the world, and he now addresses the question of occult powers at work in it. He says that there are three forms of causality in the world: first, there are interactions between two bodies; second, there is the "influx," which affects a body but is itself not of bodily origin—it is essentially occult; third, there is the "afflatus," a force which neither originates in a body nor affects a body, but rather, acts upon the soul.

"The influx emanates from an immortal body, the afflatus from an immortal incorporeal source. The afflatus imparts knowledge, the influx acts as a motive force." This means, then, that the influx is, after all, of bodily origin—it proceeds from the celestial bodies. This occult effect was the subject of astrological investigations. It is probably implied in Cardano's statement that the celestial bodies are not "ordinary bodies." The "afflatus," or inspiration, as we would say today, comes from God.

Appropriately, the discourse now focuses on Him as the Creator. But God is more than that. He is highest and immeasurable perfection, reflecting only upon Himself. He is all light, He is motionless and unchanging. No mortal can endure His light. It would be easier to look into the midday sun of summer for an entire year than to let one's mind face the divine light for even the briefest moment. Still, anyone on whom this experience might be bestowed would at that moment be the most fortunate of men. But only the good and wise may experience such ecstacy.

Like the intellect, God is always at rest; however, He is not Himself intellect, but rather, something far superior. No one knows what God is, except God himself. For this reason we should not describe God with a name, because names refer back to natural phenomena, to forces and attributes with which man is familiar. The divine souls and their essence are incomprehensible to us, but the nature of God, the most Exalted, is so to a far greater extent. Consequently, deities have no names, and man should not give them any, either.

The book *De Subtilitate* concludes with the praise of God as the source of all good.

As I already pointed out in my discussion of *De Natura,*

Cardano, departing from Stoic doctrine, differentiates between heat and the World-soul. This allows him to treat the concepts of soul and mind in the same way Plotinus did—whom, as we saw, Cardano regarded as a man of almost divine wisdom. Plotinus's doctrine of the soul is a commentary or an amplification of the concept of the World-soul as put forth in Plato's *Timaios.* Related to this doctrine is that of "Time and Eternity,"[15] which Cardano discusses as well.

He gives a new and peculiar turn to the idea of "aevum" by designating it as a "space" that the intelligences conceive of and in which they permanently dwell. At the same time, the cosmos also rests within the aevum. This notion introduces a spatial aspect of aevum in addition to the temporal one. Thus, it becomes an "imaginary" space[16] beyond the cosmos in which both the material and the spiritual world are contained. Cardano expresses the same idea in *De Natura,* when he describes the material as a perpetual emanation from the incorporeal, at the center of which the finite cosmos rests. Here, Cardano's ideas point in a direction which subsequently assumed great significance, and which eventually led to Newton's conception of absolute space. Cardano's formulations are still somewhat uncertain.

He basically wants to communicate the vision of an idea, and we should, therefore, not take his expressions too literally. After all, the aevum is another world and cannot really be described.

I would like to interpret the concept of aevum as a metaphor for the collective unconscious in the Jungian sense. It is the domain of mental powers, of those unknowns we call psychic forces. It is the source of occult workings, hidden motives, and inspiration. From it flows time, which might be called the "place" of our consciousness. The aevum, which contains our consciousness, our world, may be viewed as a metaphor for the unconscious, and the world as a representation of our consciousness. Cardano's belief that all things are animate supports such an interpretation. The soul is projected onto the world. It is the cause of order, since it is the soul from which those organizing principles stem by which we orient ourselves in the world.

In Cardano's view, God is not contained within the aevum. He is absolutely transcendent. And He is not merely the highest intelligence—He is much more. In order to convey his conception of God, Cardano refers to I Tim., 6:16: "[W]ho only hath immortality, dwelling in the light which no man can approach unto; whom no man hath seen nor can see. . . ." In his *Hymnus*[17], he says of God:

Ita praesens es omnibus, ut nullibi sis: ita magnus, ut nihil extra te sit.

Qui nullo in loco es, sed ante omnem locum in temetipso solum: non magnus, non parvus, sed immensus. Instituisti vitas has etiam ante locum, ante tempus in semetipsis, atque exinde in te solo constitutas, mundum vero et locum, ac tempus continentes, ut quod sit solum esse velis.

You are thus present in all things that You are nowhere: You are so vast that nothing exists outside of You.

You are not in any one place but are beyond any place, alone in Yourself: neither large nor small, but immeasurable. Those beings (the angels) which You created outside of space and time to exist in You alone, nonetheless contain the world, space and time, in order that all which is shall be according to Your will.

These passages may serve to supplement what Cardano said in *De Subtilitate.* His statements sound genuine; he is sincere. The assertion often made in later times, that Cardano was basically an atheist, lacks credibility.

As a theologian, Cardano follows neo-platonic-Christian tradition, which was newly inspired during the Renaissance by the writings of Dionysius the Areopagite. Nicholas of Cusa held similar views.[18] He too concludes from Paul's letter to Timothy, where it is said that God dwells in inaccessible light, that God is undefinable.

I have attempted to describe the work *De Subtilitate.* Cardano did have models, and he mentions them: Pliny and Albertus. Their works became exemplary for other books of encyclopedic scope, such as the *Margarita Philosophica* of the Cartusian Gregorius Reich (1467–1525), which soon became widely known and continued to be read throughout the seventeenth century.

Compared to these earlier authors, Cardano's philosophy is much more original. In addition, he offers a wealth of new or hardly known facts. At the same time, he has to review established knowledge, but he does so quite critically, and he never loses himself in stories of an obviously fantastic nature. Often, he speaks from experience, and it is then that he best demonstrates his great knowledge of the world and his keen insight into human nature. His manner of presentation certainly lacks coherence, and it often goes to extremes, but by these same tokens it is also most lively.

The universe, as Cardano conceived it, exhibits a graded order, and this order is purposive. Consequently, we encounter the notion that this order might correspond to varying degrees of evolution. Since Cardano believed that the world is governed by uniform principles, his thoughts easily wandered from one topic to another, which makes his presentation seem very disorganized. He himself probably did not perceive it that way. To give an example: when he is about to talk of man and his longevity, he immediately thinks of bees, since they are both of delicate build and long-lived. In the same context he mentions tortoises, since they too are long-lived, and only then does he get to his real topic, the delicateness of the human being. This leads him to relate the rather unsavory story about the corpse of Alexander Medici, who was murdered in 1537. The story is intended to illustrate the extraordinary fineness of the human body. The concluding remark that delicateness brings with it longevity as well as distinction seems illogical at first. But one has to remember that not only man is a particularly noble creature, but that the bee is equally distinguished: "esse apibus partem divinae mentis et haustus aetherios dixere" ("it is said that the bees partake in the divine spirit, in the heavenly breath," (Georgica IV, 220) is a famous verse by Virgil which Cardano quotes in *De Rerum Varietate* when he discusses the concept "mens."

Cardano wanted to present to the reader a philosophical-scientific view of the world. To this end, all particulars were shown in their interrelations and could, therefore, rarely be discussed in great detail. Cardano tends to point things out and then merely suggest possible explanations. He obviously assumed that the educated reader would gather from this what was meant and would understand what was essential. The less knowledgeable reader would not look for profundity anyway, but would still gain a general impression and receive varied stimulation. Cardano's descriptions of mathematics at the beginning of chapter XVI, *De Scienciis,* for example, are intelligible only to mathematicians, but they still impart on the general reader some idea of the nature of the subject. Cardano's writings on philosophy and science aim at being comprehensive and broad. They are intended to reach anyone who has an interest in learning. With his book *De Subtilitate,* Cardano succeeded in this most admirably.

Modern science has again turned toward the particular and narrowly defined areas of specialization. An experimental-mathematical method has been developed which, precisely because of its limited scope, has been extraordinarily successful. But with it a view of the world was relinquished which had encompassed heaven and earth, the spiritual and the physical.

5 *Astrology*

W*HILE THE BELIEF* in dreams and other omens was regarded by many to be mere superstition, astrology, in the sixteenth century, was a respectable science. Astrological predictions have, of course, always been questioned by individuals—be it for scientific or religious reasons—but at that time the critics were in the minority.

The belief in this science was particularly widespread in the academic community. Like Cardano, physicians educated at the universities frequently were also astrologers. Astrology was actually an auxiliary science to medicine. It was believed that the "geniture"—the constellation of the stars at the moment of birth—determined a person's physical and mental constitution. An experienced observer of human nature could indeed arrive at remarkably accurate character sketches on the basis of such information.[1] Astrology was also used as an aid in medical prognosis. The significant factor was the position of the moon relative to the planets at the time of the onset of an illness. From following changes in the constellation, which could be calculated, inferences were made as to the future course of the illness. This offered the chance to employ appropriate medical counter-measures.

Advocates of astrology could rightly point out that their prognostications were no less reliable than medical prognoses. Cardano, who was highly esteemed as an astrologer, considered astrology as part of prescience, and therefore not as true science. He likened its relation to philosophy as a whole to that of the prognostic practices of Hippocrates or Galen to the science of medicine as a whole.[2]

In any case, astrology was a recognized academic discipline. Until 1572, the University of Bologna had at least one professorship in astrology.[3] When the first chair for astronomy—"The Savilian Chair"—was established at Oxford in 1619, astrology was an integral part of the curriculum. The statutes of 1636 still say that a professor must explain and teach "totius in universum divinatricis Astrologiae."[4]

Aside from Ptolemy, the foremost authorities were the Arabian scholars Albumasar (died 886), and Albohazen Haly (thirteenth century). Their books were already published in the fifteenth century, for example, Haly's *Liber in iudiciis astrorum,* Venice 1485. Copernicus also owned a copy of this book.[5]

Yet, it was just the "Copernican revolution," his heliocentric world-system, which discredited astrology. As long as one could divide the universe into superlunary and sublunary realms, it was plausible that the heavenly bodies above the moon were of a more noble substance. While the laws of growth and decay governed the sublunar world, eternal principles were thought to rule the heavens; the celestial bodies were obviously spiritual beings. In the heliocentric system, however, earth itself became a heavenly body, which made a differentiation between the sublunar and the celestial world untenable. Initially, the belief in astrology was not shaken by the new scientific doctrine. Yet Kepler had to develop his own theory in order to reconcile his heliocentric conviction with his belief in astrology. He based his theory mainly on his light-metaphysics and on the concept of the divine significance of classical geometry, where only the construction of certain angles is possible which, therefore, produce an occult effect upon the soul.

Cardano did not reject the Copernican theory completely. He even considered the notion that the moon alone revolves around the earth rather good judgment, "since its effects are indeed quite different from those of the other planets."[6]

It will be easier to appreciate that era's belief in astrology if one realizes how incredibly precarious life was in those days. People were defenseless against natural disasters, wars and epidemics, the caprices of those in power, as well as the snares of personal enemies. But whatever one's fate, it was ordained by God, and had even been forecast by the signs of the heavenly bodies, for our comfort and our salvation.

Man ought, however, to submit to the inevitable. Therefore, Cardano was remarkably unperturbed whenever his predictions proved wrong. This did not invalidate astrology, it merely showed how difficult it was to interpret the signs and how easily a significant circumstance could be overlooked. Besides, it might have been God's will that the astrologer should have erred, since a correct prognosis might have put him in still greater danger. Only after events had actually occurred could one see clearly that they had been predicted by the stars. Newton held very similar views with regard to biblical prophecies. He interprets the prophecies of the Apocalypse of Daniel as relating to historical events of the Early Middle Ages, whereas he sees the Apocalypse of St. John as pointing toward a later time. He adds: "God gave this and the prophecies of the Old Testament not to gratify men's curiosities by enabling them to foreknow things, but so that, after they were fulfilled, they might be interpreted by the event, and His own providence, not the interpreters', be then manifested to the world."[7] Being a sensible man, Newton judiciously refrained from using the Bible to predict the future. Cardano attempted such predictions again and again with the aid of astrology, never becoming discouraged by failures which he admits openly. For "the heavens are the instrument of the supreme God through which he effects, advances, and governs all that occurs on earth."[8] Of course, "anyone not versed in this art will soon get caught up in ambiguity, complexity, and fantastic speculation."[9]

Since I am not versed in this art and would rather not fall into fantastic speculation it might be best to illustrate Cardano's astrological ingenuity with three examples. In his *Liber de exemplis geniturarum,* Cardano compiled one hundred horoscopes, mostly of notable personalities—Emperor Charles V, the French king François I, the English king Henry VIII—but also those of cities, Florence and Bologna, for example. The first of these horoscopes is that of Petrarch, the hundredth that of Albrecht Dürer. The horoscope of Erasmus of Rotterdam is the twelfth. These are the horoscopes I should like to present here, since they show how these men, whom Cardano greatly admired, appeared to him in the light of their nativities. They will also give the reader an idea of the manner in which Cardano interpreted the signs of the heavenly bodies.[10]

1. *Petrarch*

In position with Jupiter are small stars having the nature of Saturn and Venus; five small stars of the nature of Saturn accompany Mars in Gemini.

With Venus is the star in the foot of Gemini—being of the third magnitude and having the nature of Mercury, not of Venus. With Mercury and the ascendant appear small stars of the nature of Saturn and Mercury. With the ascendant rises Sirius. In the depth of the firmament stands the spike of Virgo. She has the nature of Venus, not of Mercury. In addition, there are Arcturus and Boetes, which have the nature of Jupiter. They are of the first magnitude. This remarkable man had many extraordinary traits.

First: the artistic sweetness of his verse which is celebrated throughout the world. This is granted by Jupiter in the house of Mercury, in quartile to Venus at the foot of Gemini, which has the nature of Mercury and not of Venus. Venus is, however, in trinal to the moon. But the moon's power is lessened by the rays of Jupiter who is in opposition. This signifies perfect euphony.

Second: depth of feeling combined with the highest degree of diligence. This is granted by Mercury in the ascendant beside the Sun, in sextile to Saturn. It is greatly intensified by Sirius who is rising in the ascendant and who has the nature of Mercury. Thus, vigor was added to artistry and eloquence.

A third fact is his lasting and undiminishing fame, accounting for translation of his work even into Spanish. This is signified by the spike appearing in the depth of the firmament.

Fourth: he remained a bachelor and had no offspring. This is indicated by the Sun in quartile and Mercury in angle position facing the tenth house. But they stand in barren signs, as does Jupiter, who is in quartile to the fifth house. Also, Saturn, shining in sextile, is in an infertile sign. Accordingly, Petrarch only had one son whose unfortunate destiny was indicated by Jupiter—he lived for only a short time.

Fifth: Mercury granted him the gift for language, he excelled in both Latin and Greek.[11] A powerful Mercury always grants an aptitude for languages, particularly if he is in close position to the Sun. The Sun in the east signifies long life. Jupiter in the second house furthers religiosity, something he does, incidentally, in most other positions as well.

12. *Erasmus*

Now I come for the first time to you, my Erasmus, flower of our age. You are that much smaller in stature than Cicero as the Roman Empire is larger than the German. And as eloquence comes more naturally to an Italian than to a Barbarian, it is also more easily attained in one's mother tongue than in a foreign one. Despite the unpropitiousness of the heavens you dedicated yourself to sweetness of speech, and barbarism's indebtedness to you is far greater than the detriment it presented to you.

In conjunction with Mercury and Venus the spike of Virgo emitted its brilliant rays. And the moon in its node promoted keenness of mind and continuous good fortune, for it stood in the nativity. We have often been able to observe this. And to all this Jupiter was favorable, who shone out of the ninth house in the sign of Gemini, emitting its rays in trinal. As a result of this Erasmus was, so to speak, imbued with spiritual purity; hence, he did not strive for worldly honors commensurate with his talents[12], but aspired instead to the mastery of languages and—in order that the perfection of the number seven be fulfilled—to a thorough knowledge of the Holy Scriptures. Mars stood in opposition to Jupiter, which caused illnesses to recur in the seventh house: thus Erasmus was repeatedly afflicted with calculi. Saturn in the seventh house signifies great danger and the frequent snares of enemies. The node of the moon being in conjunction with the sun—they are only separated by two degrees—denies him wife and child. Its position in Scorpio also leaves certain secrets unrevealed. This lends even greater credence to our calculations.

He lived to the age of seventy, and he was fortunate in that. For he had neither too little time to pursue his studies, nor too much of it to expend on pedantry and frivolities. For such a man I could not think of a more favorable geniture, even if I could choose it myself. The sun, the moon, Mercury and Venus are free of portentous radiation, while Venus, joined in the nativity to the spike of Virgo, indicates his exceptionally subtle humor as well as furthering his success. He died when Venus, the lady over the figure of the nativity and the ruler over life, came into opposition to Jupiter and into quartile to Mars.

However, not everyone on whom such a geniture is bestowed will equal Erasmus. As I mentioned elsewhere, propitious revolutions of the stars must follow, as was the case with Erasmus.

100. *Albrecht Dürer, the painter*

He excelled in the graphic arts and left testimonials to his art in books as well.

The moon with its node and the sun were in quartile to Jupiter; the sun and Venus were, however, in sextile to Mars. Moreover, Venus, the sun, and the moon were in the constellation of six magnificent stars, four of which shine in Orion and two in Auriga. Three of these are of the first and the others of the second magnitude, making this the most luminous place in the sky. The great power of these stars combined at the time with that of the planets. Whether this exerts a greater influence on painting, on literary work or on other artistic pursuits, whether the nature of the stars or the celestial region is of greater significance here, still needs to be investigated. He died in 1528.

It is not surprising that Cardano begins his discussion with Petrarch, since he was—next to Luigi Pulci—his favorite poet.[13] It is astonishing, though, that he concludes the series with the German painter Albrecht Dürer, the only graphic artist in the group. Of course, Dürer's magnificent engravings were greatly admired in Italy, and he was also a scholar and a writer, which must have particularly impressed Cardano.

But Cardano speaks of no one with the same affectionate tone in which he discusses Erasmus, whom he venerates almost like a saint. He admires him as the Christian humanist who combined the sincerity of his religious faith with friendly humor.

Within the astrological context it is notable that great significance is not only attributed to the planets, but also to the fixed stars, and that besides the signs which are measured from the vernal point, the actual constellations are taken into account.

6 *The Interpretation of Dreams*

IN HIS WRITINGS on astrology Cardano for the most part followed the classical authorities. As an interpreter of dreams, however, he is of remarkable originality. Antiquity left us no substantial theoretical work on this art which claims to be scientific and which could, therefore, have served Cardano as a model. It is unlikely that such a work existed. Cardano tried to fill this gap with his *Synesiorum Somniorum omnis generis insomnia explicantes, libri IV.* According to a note in *De Libris propriis,* he wrote the book around 1545, but later revised and enlarged it.[1] It was published in Basel in 1562 and translated into German the following year.[2]

The point of departure for Cardano's theoretical considerations is the book by Synesius of Cyrene.[3] He lived around 400 A.D., he was a neo-Platonist and pupil of Hypatia at Alexandria. Later on in life, he converted to Christianity and was ordained as bishop. Synesius seems to have been an amiable and judicious man who was free of all worldly ambition. He therefore found his episcopal office rather burdensome.

His *Book on Dreams* is a short treatise—he claims to have written it in a single night. In it he defends the merits of dream interpretation and offers a philosophical explanation for his views. Central to his theory is the idea that the universe is a unified and animated entity. Yet this unity is not a simple oneness; it is composed of many parts. Consequently, as in all relationships, discord exists alongside harmony. The universal interrelations of phenomena provide the basis for the art of divination. This art opens the path to God and makes possible the ascension of the soul. In the pursuit of this goal divination by dreams is particularly useful. All that is required to practice it is to go to sleep after having washed one's hands and said one's prayers. Every person can dream, whether rich or poor, master or slave. The dream is a democratic and philanthropic phenomenon.

There are sublime dreams in which we experience a union with God; these need no interpretation. But most dreams are obscure and have to be methodically interpreted. How can one master the art of interpretation? First, one must sharpen one's powers of perception. Philosophy is quite useful for this because it frees one from the interference of emotions. There is no general method of teaching the interpretation of dreams. It is an excellent idea to write down one's dreams—to keep, so to speak, a "night-diary." This is also a good exercise in the proper use of language.

These are the basic general considerations with which Cardano starts. In analogy to our calling modern dream analysts either "Freudians" or "Jungians," we would call Cardano a "Synesian," as expressed by the title of his book. It is dedicated to Carlo Borromeo, Cardinal-Archbishop of Milan, whose family had greatly benefited from Cardano's superior abilities as a physician, and who had become his patron. Borromeo, who was eventually canonized, was one of the cardinals who defended Cardano in his conflict with the Inquisition. The preface of the book is addressed to the cardinal. In it, Cardano asserts that the interpretation of dreams is as much concerned with the future as is the work of the physician, or, for that matter, the farmer or the military leader. He concedes that one might perform this art poorly by drawing conclusions too readily, and that one might often arrive at the wrong conclusion. But then, a physician might kill a patient, a judge arrive at the wrong verdict, a barber cut a customer's throat. It has often been suggested that these practices are less open to error than is the interpretation of dreams—particularly when the latter is practiced by the general populace—but this is not so: all paths can lead us astray; varied and ever-present is the possibility of error; we are bound by our ignorance. There is, however, one safeguard against error: to acknowledge that all phenomena are natural.

When the gable of a house caves in, or when a dog howls, it is a natural occurrence. Everything depends on the frame of mind of the person confronting such occurrences. Dreams, too, if they have any meaning at all, signify some natural occurrence. Moreover, whatever is revealed by them can be prevented in many different ways. It is only when inevitability and divine ordinance are ascribed to events in dreams that superstition takes over. Dream revelations should, therefore, not be looked at with superstitious awe, nor should they be anxiously heeded as divine precepts. Cardano suggests that dreams may be less indicative of future events than they are of the dreamer's present situation. From such insight into the present the dreamer should nevertheless make inferences about the future. It is apparent that such inferences about the future can be made from the most natural occurrences. Of course, it is dangerous to rely on conjecture alone. Yet, do we not allow most of our actions to be guided by conjecture? What is important is that this be done with caution and circumspection. People who believe in dreams, Cardano says, have never been known to act on impulse. On the other hand, many people, like those who make use of astrology, for example, rely too much on their own abilities, strengths, and craftiness, and they therefore disregard the warnings signaled by dreams.

Cardano's work is divided into four books, the first and the last of which are by far the most extensive and the most interesting. In the second chapter of the first book Cardano says: "An inquiry into the phenomena of dreams must take three basic factors into account: the dream images, the dreamer's way of life, and his individual character and disposition. Accordingly, the treatise is divided into three books; a fourth one contains selected illustrative examples." The first book does, however, offer much more than an explanation of "dream images." It presents a complete theory of dream interpretation and constitutes the principle part of the whole treatise. In addition, the book includes in its more than fifty chapters an encyclopedia of dream symbolism, arranged in correspondence to the order of the universe. Cardano employs traditional concepts used by professional interpreters of dreams since Artemidorus, and one might well presume that this part of the book is obsolete today. However, this is not the case. Dream thoughts are basically archaic in character; thus, they are akin to the modes of thought of antiquity and the Middle Ages, when everything still carried a hidden meaning. It must also be noted that Cardano himself had an extraordinary awareness of his unconscious, as evidenced by the account of his own dreams in the last book of the work, and he was obviously well-qualified to make pertinent observations on what I shall call "dream language."

Anyone can discover that dreaming—like any other activity—will benefit from training and cultivation. A person who observes, writes down and examines his dreams will find that after awhile the dreams will become ordered, more intelligible, and more interesting. I don't believe that this is due only to clearer observation and recollection of the dream. Rather, I think that the activity of our unconscious imagination—and this is what dreams are—gradually becomes more expressive and approaches the state of consciousness, thus making an intelligible impression on the latter.[4] This establishes a relationship between consciousness and unconscious dream work. The kind of ordered form and meaning that dreams then assume depends, of course, on the individual consciousness that analyzes the dreams. The most expressive dreams are not only characteristic of the individual dreamer, but also of his time and its collective consciousness. In this sense, dream language is a reflection of a particular era and cultural tradition.

Synesius's remark that the notation of dreams cultivates the proper use of language is not, as the editor seems to imply, simply rhetorical.

On occasion, Cardano's interpretations undoubtedly appear strange. One should keep in mind, however, that the dream language of his time was different from ours, and that Cardano's entire view of life was highly "symbolic."

Report on the Content of the "Synesiorum somniorum libri" (Book I)

Cardano begins with a brief review of the existing literature on dream interpretation. The authors include

Artemidorus, Solomon the Jew, Nicephoros Gregoras—he wrote a commentary on Synesius—as well as Persian, Egyptian and Indian writers. Artemidorus and the Hebrew passed on a number of valuable observations, but their investigations were unmethodical and their assertions often incorrect. Synesius stressed the significance of dreams, but he wrote nothing about the technique of interpretation. Cardano intends to make up for this lack of information.

Types of Dreams and Their Sources

Dreams have either physical or psychic causes. Food and drink and the bodily humors—black and yellow bile, blood and phlegm—act as physical stimuli. Psychic stimuli are anxieties, ideas, memories and emotions. Dreams can also be caused by divine inspiration. Dreams which are stimulated by the humors of the body and those of supernatural origin are especially prophetic.

All of these factors produce a tremor of the animal spirits which takes on the form of a dream. The dream-material is made up of images derived from memory. In the second chapter of the book Cardano uses the term "transposed" memory to refer to incomplete images. It would apply, for example, to a dream situation in which one sees a complete stranger or some unusual, unknown plant. It is not until the fourteenth chapter, however, that Cardano explains "transposition." To understand it, one must be aware that dream images are related to one another by association. Images blend together, and one image can transform itself into another. Thus the thought patterns of the dream state (ratio somniantis) are formed. When a person dreams of something which—if taken literally—is totally out of character, the interpretation will require a transposition of the material onto facts or circumstances that can most reasonably be associated with it. Let us suppose a person dreams that he is cutting someone's throat. But the dreamer is not a man of the sword at all, but rather a man of the word. It appears, however, that he tends to evade facing superior arguments of an opponent by cutting off debate, or that he has just refuted a book someone else has published. To take another example: someone sees himself despairing in prison. This means that the person is engaged in such difficult studies that life has become torture for him. Cardano uses this same approach when he interprets strangers as representing the dreamer himself, who, because of emotional troubles, does not really know himself. In the sixty-fifth chapter Cardano says explicitly that such strangers signify psychological perturbations (perturbationes animi). We must assume that these are alien to us. Although they affect us most intimately, we deal with them as we do with strangers.

These are some examples of "transposition." If it is true that all dreams originate in memory images, it would be most interesting to know what newborns dream. They do not have memories of actual reality, of course. Should they nonetheless have dreams, we would have proof of our spiritual immortality.

The form a dream takes is also an indication of its source. Dreams that are stimulated by food and drink are generally confused and fragmentary. In most cases, they are also short, since foods do not cause a prolonged agitation of the humors. Only the humors produce rather steady, ordered and lengthy dreams composed of a variety of remembrances. The content usually reflects experiences and thoughts of the previous day, although the recollections may also stretch further back in time. Dreams of this type recur frequently since the nature of the humors remains generally unchanged. True memory dreams are usually induced by strong emotions but have little psychological impact.

Cardano believed that memory images were localized in the brain, rather than in the vital spirits. He points out that a temporary loss of consciousness cannot obliterate them, whereas damage to the brain could completely destroy them. Dreams of superhuman origin are very soothing; they do not agitate the dreamer, but they nonetheless make a deep impression on him. Such dreams have higher causes, that is to say they are due to the actions of the heavenly bodies.

Whatever the cause of the dream may be—even if it is of higher inspiration—a great deal depends on whether the dreamer commands an adequate diversity of images to enable the psyche to accurately represent the dream material. The more unusual the material is, the more difficult it will be to have the proper dream. Dreams which originate in the celestial realm are produced by the effect of the "influxus siderorum" upon the vital spirits. This effect can, however, be interfered with by such factors as indigestion or emotional upsets; then the dream becomes incoherent. Without such interference, these dreams are intrinsically intelligible and most often rather brief. They are the true prophetic dreams.

The Veracity of Dreams (Fourth and Fifth chapters)

To have veracious, that is to say meaningful, dreams, it is necessary to lead a well-ordered life, to devote oneself to important tasks, and not let oneself be upset by fear or sorrow. If a person is truthful and devout, he will also have truthful dreams. Meditation greatly contributes to having orderly dreams.

Then there are seasonal influences to consider: dreams will be more veracious in winter and summer, when the weather is less variable than during spring and fall.

Dreams which form a sequence are truthful and of particular significance. The longer the time span that separates them, the more reliable their message.

People who seldom dream usually have significant dreams. One's dreams are most significant in youth and in old age.

Principles and Rules for the Interpretation of Dreams

Since almost all dreams are produced by an interplay of diverse factors, and since even those of superhuman origin are often distorted by the effects of the body's humors or by emotions, dreams generally need to be analyzed to become intelligible. The interpretation of dreams proceeds from the idea that similar things harmonize, whereas dissimilar things conflict. What is natural for a king or a young man is also good for them. But the same thing is not good for an old man—like Cardano considered himself. To recognize what is natural for a person, one has to take many factors into consideration: temperament, sex, age, place of origin, ancestry, rules of conduct, religion, rank, possessions, children, health, friends, family, spouse, profession, esteem, manners, scholarship, and habits. Obviously, the interpreter must obtain comprehensive and intimate information about the candidate and his family in order to judge what would be in character in a particular instance.

For example: an old man dancing signifies folly, indecency, or death. It means death because it reminds one of the "dance macabre."

All interpretations must take transposition into account: a motive may at first seem out of character, but by association it can be transposed onto an emotion or an object that is closely connected with it. First, the various general connotations of the dream imagery must be examined. If these point to a common source that relates to the facts or circumstances of the dreamer's waking life, the dream imagery becomes intelligible. The emotions aroused by a dream are also important. If the dream refers to the present but does not evoke strong emotions, it pertains to a matter of minor importance. If, on the other hand, the dream concerns either present or future and evokes any emotion at all, then the dream reveals matters of significance. The interpretation of dream symbols is based on the analogy of microcosm and macrocosm, of man and the universe. One must, of course, always keep in mind that one thing can have several meanings. For instance: a horse alone can signify a spouse; a horse with a rider signifies flight; ownership of a horse represents wealth. The correspondences between the microcosm of the human body and the macrocosm of the universe are specified in several columns. Following is an excerpt from this tabulation:

The head: master, house, king, mind, honor.

The hair: roofing tiles, ornament, grief, bodily harm.

The neck: the spouse (since she graces and maintains the house), pleasure and vigor.

The arms: right is masculine, left feminine.

Several things can also symbolize different aspects of

one subject. Thus the heart, the eyes, the penis, and the branch all symbolize the son. He is represented by:

The heart, because he is wise and loves his father.

The eyes, because he is amiable.

The genitals, because he is unchaste, perhaps even illegitimate.

The branch, because he is dull and useless.

I have so far discussed Cardano's general theory of the dream life and its proper interpretation. This corresponds to the first fifteen chapters of the first book. The remaining chapters comprise what I have called an "encyclopedia of dream symbolism." Beginning with the deities and the celestial realm, it moves downward through the elements to the profusion of worldly phenomena. I would at least like to give a few examples of how Cardano did this.

The *deities* are in this context considered part of the worldly sphere. They usually signify general causes, often of unfavorable portent.

Demons and *heroes* signify foremost our emotions. Heroes have a more positive meaning than demons, because they challenge us to be virtuous.

A dead person—regardless of whether he appears dead or alive—signifies death, especially if he is calling and one recognizes his voice but cannot see him. Gifts received from the dead have a positive meaning. Seeing one's parents dead is not particularly alarming, especially if they are actually dead. An important point is that it is more ominous to see the deceased dead than to see them alive.

The *sky* and the *stars:* since the sky is common to all,

it symbolizes one's homeland[5] or fate. The stars signify any kind of corporation, a council, the community.

Air, not felt yet everywhere, signifies the heart and the chest, as they are always in motion. It is also a symbol of religion and of life. Cloudy air signifies religious strife. A house in the fog is a symbol of vain hopes.

Water: rivers are symbols of the course of our lives. Drinking from them signifies death. In the case of scholars, however, it can also signify almost divine wisdom. As the river—so life and wisdom. Water in all forms has a naturally positive meaning.

Earth symbolizes the intestines, furthermore the mother and the native land. When the earth speaks it signifies long life; it speaks the truth.

Mountains and *forests:* mountains signify powerful men or difficult business transactions. Forests harbor danger (robbers) and signify onerous chores.

Physical activities: flying represents hope. Flying toward heaven can mean a glorious death. Moving about in great leaps and bounds signifies vain efforts.

Fishing, hunting, bird-catching mean deception or a battle, holding good prospects and little danger.

Drinking wine expresses confidence and serenity; when one gets drunk, one feels able to overcome all difficulties.

Riding a mule alludes to a boring enterprise, involving little risk but also holding little hope of profit; this is indicated by the animal's smooth and sure gait.

These are some examples of the way in which Cardano interprets dream symbols. I am sure the reader will appreciate the poetic truth of these interpretations written more than four hundred years ago.

Rules for the Interpretation of Dreams

In chapter fifteen Cardano enumerates sixty-three basic rules for the interpretation of dreams. He says:

Honest people have premonitory dreams. The dream content may refer to future events just as often as to things past or present. Every dream can be reduced to its general structure (ad sua generalia), which will reveal its essential meaning. For example, if I see a mountain sliding down on top of me, this mountain signifies something big. Institutions and people in power are big. Or, someone who is seen falling down will be overthrown, stripped of his position or rank. In this way one can make conjectural judgments. Every dream, even the most menacing, can be beneficial if it is a wish-fulfillment. A case in point would be if someone who is eager for revenge sees a bloody sun.[6]

Whenever something which is generally thought of as singular, such as the sun, the moon, or—in a Christian society—the spouse, appears in multiple form it is a sign of strife and disorder.

It is very important to observe the sequence of dream phenomena, to see how one thing merges into the next, and how it all finally ends.

It is easy to make mistakes if one does not keep in mind that in dreams one thing can have several meanings. For example, if a person sees himself as having three eyes, it means that he is clever and that he will be successful. But it can also mean that intelligence must be applied as failure is imminent.

Dreams about things very dear to us are seldom propitious, as they usually arouse strong emotions, and most emotional agitation is harmful. The image of something momentous is always a bad omen. Good things are never represented by something large; inordinate size usually indicates a disaster, dreadful events like the plague or the collapse of one's house. A distinction must be made between the natural meaning of things and the meaning assigned to them by convention and custom. I am thinking, for instance, of the love for an illegitimate son or a friend. Dreams tend to turn from the general to the particular. A physician stands for your own physician. On the other hand, your own physician can stand for physicians in general.

Another important aspect of proper interpretation is the "Reciprocatio." Dreaming of a stranger, for example, may mean that this person does not know you. Anything that is quite obviously not a reference to our own person—like deities or the dead—can, however, not be interpreted in terms of "Reciprocatio."

Book IV: Examples

The book *De Exemplis* begins with Cardano's interpretation of dreams transmitted to us from antiquity, dreams of Alexander, Caesar and others. Here, Cardano for the most part conforms to traditional interpretation. He then goes on to discuss some dreams related to him by acquaintances. Most of the examples are, however, furnished by his own dreams. Unlike Freud, Cardano was not motivated by professional discretion in the choice of his material, but rather, by his theory that the dreamer himself is the most qualified interpreter of his dreams. Only he can recall all the details of the dream and the circumstances under which it occurred.

Cardano begins by differentiating between two types of dreams. He says:

> One kind represents the things themselves, these are called "Idola." They usually consist of conversations which one hears, rather than of things one sees. The purpose of such dreams is to confirm an opinion about a particular problem. Then there are dreams which express a natural symbolism and random thoughts and considerations. Such dreams present the real test of interpretative skill. The details of this art cannot be taught

by providing a set of rules. The best way to illustrate the basic technique is by sample interpretation. A special aptitude is indispensable, but even then dream interpretation remains a matter of some uncertainty.

The first example is a dream of one of Cardano's friends, interpreted by Cardano. He notes:

A friend of mine, a rather ignorant fellow, was supposed to take an exam. He was greatly troubled and agitated by this prospect. The day before the exam he dreamt that he was going hunting through fields, accompanied by friends and by his hounds. Suddenly, a band of robbers appeared, taking all his friends prisoner except one. This one seemed to be a friend of the bandits. The dreamer himself had hidden in a cornfield and was thus not discovered. But he was afraid that his dogs might betray his hiding place.

The following morning he came to me, filled with anxiety. I told him: Cheer up! The bandits signify the danger of making mistakes. The hounds that might betray you stand for the exam itself. And as to the friend of the bandits—that is me, because I am not afraid of making mistakes. The cornfield where you were hiding symbolizes the things you have retained in your memory. After hearing this explanation, my friend left in great astonishment.

Cardano's interpretation is literal and simple. He closely observes the dream imagery and the actual circumstances of the dreamer, and the way in which he makes himself part of the dream situation is a sign of his true sympathy for his friend's distress. The interpretation strikes me as an ingenious therapeutic means to get the candidate out of his state of panic and to reassure him. Following are some of Cardano's own dreams:

1. On the sixteenth of February 1540, I believed myself to be lying in bed with a black sun glowing above me. The rays

were brilliant yet blacker than ink. I thought it was the sixth of April, the date for which an actual eclipse of the sun had been forecast. To my surprise I could recognize stars, shining as they do in the night sky. Upon awakening I realized that this was a portentous dream, but I could not decipher its meaning, partly because I had not yet reached the necessary level of insight in the art of dream interpretation, and partly also because fate would have it otherwise. I know now that this solar eclipse signified the death of my son. At the time of his actual death, however, he had already been accepted into the College of Physicians, an event which in 1540 still lay in the distant future. The darkness which signified his death let stars (symbols of ordinary physicians) become noticeable that would have remained hidden otherwise. The fact that I thought the date was the sixth of April portends to an act of total destruction, not just an ordinary death or some future banishment.

This dream is interpreted as a "vaticinium ex eventu." It shows how Cardano tried to come to terms with the execution of the son in whom he had set such high hopes, by viewing the catastrophe as fateful and preordained. He was forewarned of it, but fate did not allow him to understand the premonitory meaning of the dream. In the next dream, he again recognizes allusions to this, his most grievous loss:

2. I have written down a most wonderful dream, and only now while I am writing this do I understand its true meaning. Had I been able to understand it initially, it would have forewarned me of everything that has since happened to me. It is an excellent example of a dream of revelation (somnium specularis).

Because of its mysterious nature and the subtleness of approach necessary to decode it, this is the most impressive type of dream. It will be helpful to first familiarize the reader with my personal circumstances at the time the dream occurred. Otherwise my interpretation will be incomprehensible. It was the year of our Lord 1544 in the early hours of January nineteenth. At that time I had two sons. The elder was in his tenth year; the younger was born the year I started my medical practice in Milan.

In my sleep I saw my friend Prosper Mario. Although I knew that he was dead, I asked him to extend his hand to me while adding "although you are dead." Upon hearing this he refused, but I persisted, and after he had actually touched my hand he calmed himself and went away. My older son followed him. When I called my son back he did not respond, and so I followed him with other members of the household. We found him unharmed in the entranceway of an isolated house. My friend Prosper I did not find again. At that moment I thought I woke up. Something behind my pillow had awakened me. I imagined being awake and reflected on my dream. I was relieved that the deceased was not threatening me or my son with any catastrophe. When I lifted my head I saw light coming through the window and was reminded that it was time to get up and attend to my professional duties. Fear caused by that something behind my pillow made me hurry. As I examined the other window to see if the light was coming from there, I noticed that the front door was open. I picked up a stick and rushed out. I was struck by surprise and fear; death stood in the doorway. Immediately I thought: this is a fatal omen unless I drive him away with the stick and try to beat him. He retreats, I follow—he flees into another house and hides in the lavatory. I saw no danger signal for myself in this, but concluded that there had to be a corpse in that lavatory. I was considering removing it when I imagined that I awoke from my second dream, although I actually went on sleeping. I did not remember anything about the previous dream, except that I thought the notion of a corpse being in that lavatory must have originated in another dream. Again I hastened to get up. Then I saw a small black scrubby dog under my bed, who in play had torn up my boots. I said to myself: I wonder who could have put the dog in here. Yesterday

I did not have a dog. I am glad I woke up before dawn, because any noise from him would have frightened me. As I was thinking this I actually woke up.

As to the meaning of all this: Prosper Mario refers to the University of Pavia; he was from Pavia, and he was a scholar. The fact that he appears as dead—which indeed he was—signifies my faded hopes of going to Pavia. For the past five years I had refused the position offered to me there. He was sad when he finally extended his hand to me; the senate, whose offers I had so far rejected, finally gave me the position almost against my will. The hand was seized; I got the position that same year. My son following Prosper means that I shall lose him upon my third return to the university. And that is exactly what happened. I resigned the professorship in 1545. I returned to Pavia in 1546, but resigned again in 1551. In 1559, I once more went to Pavia. My son died in 1560. He would not have died had I not gone to Pavia a third time, or at least he would not have died the way he did. This is signified by him entering the isolated house where my companions and I found him alone in the doorway. I did not understand this before, nor did I grasp the meaning of the three sections of this revelatory dream. Now I know that this first segment is the most important of the three. The meaning of the two subsequent dreams is less evident. Let me turn now to the second one: I wanted to get up to practice my profession; death is standing in the doorway and I drive him away with my works *De Subtilitate* and *De Rerum Varietate;* it was the death of my name which threatened. With regard to this, the dream which reveals my fate shows that my works will be lasting memorials to my name. Death hiding in the lavatory means that until my death something will remain obscured by shabby and irresolute people. So much for the second dream. The third dream is short, either because it refers to only a short time span, or because it does not want to disclose subsequent events. When I came to Pavia a third time, there was per chance a scrubby black dog in my house—he is still with me now. This can mean that there is an antagonist in the household, but that he will merely cause apprehension rather than any real harm; on the other hand it can also signify delight and entertainment. Obviously I could not be allowed to make any further inferences since I was to return to Pavia several more times. By the way: not finding Prosper Mario again seems to indicate that it is my fate to strive unrelentingly, but in vain, for success outside an academic institution.

That it was not death himself but a corpse who was hidden away in the lavatory in that third dream means that before I die I shall free myself of all anxieties and will have a chance to read my books. Of course, some worries will always remain. I shall have to be diligent until my death so that the corpse may be gotten rid of. The image of the corpse may revolt the reader, but I cannot help that. This something I felt behind my head in the second dream—the source of my fear—was my desire to write books. This desire pulled me out of my sleep, that is to say, out of an ordinary bourgeois existence, and directed me to get up and drive away death who was standing in the door.

Althouth Cardano is aware that without knowledge of his personal circumstances his interpretation of the dream must remain incomprehensible, the information he gives is sparse. Let me therefore add that Cardano was first offered a position at Pavia on probation. He did not accept. He was, after all, a renowned physician and scholar, and such an offer amounted in his eyes to an insult. In the end he was given a regular appointment, but the unfortunate circumstances of the time often prevented the university from paying its faculty's stipends. This forced Cardano to leave Pavia on several occasions to practice medicine in Milan.

The first two dreams clearly show the intense relationship Cardano had with his unconscious. He himself is active in his dream life; this means that he is trying to come to terms with whatever is happening to him. He immediately attacks death with a stick, beats him and pursues him, forcing him to hide in a lavatory. It seems logical to associate this scene with the idea of Cardano the physician fighting with death. Instead, Cardano associates the death image with the possible extinction of his name. This thought is characteristic of Renaissance mentality: everlasting fame and the perpetuation of one's name were foremost goals of the time.

3. On Ascension Day 1561, I saw myself riding a very large mule. When the sun had risen, a laughing naked woman appeared above my head and proceeded to shit and piss.

As I do not own a mule but only a small palfrey, the image signifies that I shall be given an official position in a small town, which will be onerous and useful but not particularly prestigious. This is more or less just what happened. Although there are a few undignified elements in the dream which might cause a frown, it is nonetheless gratifying to see how accurately life is predicted by it. This is so extraordinary that it is almost beyond belief.

Let me add here that in 1562 Cardano permanently resigned the professorship in Pavia and moved to Milan. He was subsequently offered a professorship in Bologna, which was at that time the secondary capital of the Pontifical State. Since the book on dreams was ready for publication by September of that year, we may assume that negotiations with the University of Bologna were still going on when Cardano wrote the above dream interpretation. For a while it was quite uncertain whether these negotiations could be concluded successfully. It required the personal intervention of Cardinal Borromeo to surmount opposition to the appointment raised by members of the university faculty.

The dream imagery is wonderfully graphic and marked by robust humor, although Cardano apologizes for the latter. The woman is obviously a representation of Fortuna or Tyche (this is at the same time a city-goddess). According to Cardano, a mule signifies onerous tasks, and he takes his interpretation from there.

4. At midday on the eighteenth of July 1561, after having had only a light lunch of some fruit and goat cheese, I saw myself calling on a senator in order to state my case in a matter of restitution of a fine of one hundred gold crowns.[7] I had hardly begun to present my argument when the senator flew into a rage and answered me in a most brusque manner. As I softly interjected a few words in my defense he became extremely agitated, squeezed my throat with his hand and pressed his elbow into my back. I humbly begged him to let go of me and to forgive me, but he only pressed with greater force and determination to the point where I actually feared for my life. There was also the danger of being beaten by his servants. At last he got tired, he let go of my battered body and threw me out of the room. I hesitated to leave the house fearing that he might throw stones at me from a window. But I did not feel safe inside the house either. I was suddenly filled with a grief much deeper even than I had felt when my son died. I also remembered having dreamt that I was led to the execution, feeling disheartened, tormented and grieved. But then I said to myself: woe is me—if people found out about this, how could I show myself in public again? If I do not bring charges against the senator now, this injustice will go unrighted. On the other hand, if I do bring charges, things will come to light of which he is ashamed. Then the entire senate will feel animosity toward me and take revenge. There is no way to predict what they could do to me should they all agree on this matter. The president of the province will not want to have any part in this either, as my case is directed against the senate. Besides, he could not do anything anyway. As I was reflecting upon my situation, an old man approached,

followed by a young man carrying a sword. I feared that he, too, might attack me, but neither of them did me any harm. Meanwhile I had left the house. In the street I chanced upon a friar whom I knew to be a very good friend of the senator. One of the senator's servants appeared as well. I did not know him, but I gathered from the conversation that this was his position. I thought it an opportune time to lodge a complaint with him and said: if the senator is ashamed of what he has done, what benevolence could he show me to make up for such an offense? At this, the servant got so angry that he seemed ready to attack me. At that moment I woke up from this unpleasant dream.

This senator stands for the senate as a whole, and as one of the prefects of the university he also signifies some matter connected with it. In actuality, one hundred crowns had been taken from me by royal writ, and I was hoping that the governor would return this sum to me. This would have been just. The dream signifies, however, that I would not get the money back. If the prefect would give it back to me, the senate would press upon my back, that is my old age, my private disgrace. I must add here, that I attacked that particular senator most vehemently when, with the aid of my friends, I filed my petition. He was one of the men who condemned my son. My emotional state was a reaction to the senator's demeanor. The whole dream sequence reveals my fear of dishonor. But since I suffered neither physical pains nor wounds, my adversaries contented themselves with letting injustice prevail.[8] The fear of leaving the house means that as long as I feared public dishonor I would not dare to resign my teaching post. And this meant that I would remain in their hands. In the end, I did emerge from the house, demonstrating thereby that I would free myself at least from fear, although not from dishonor. The dream further indicated that I could hardly expect to alleviate the burden of dishonor with the aid of friends who did not have any obligations toward me. Being seized by a feeling of grief more intense than at the death of my son points to the necessity of coming to terms with the fact that he had been a victim of injustice.

This dream illustrates dramatically the way in which Cardano dealt with the psychological and social effects which the dishonorable death of his son had upon him. It is a question of clarifying for himself a whole complex of the most intense emotions that were caused by that event. The next dream shows that he has gained some crucial insight.

5. The following day, after a lunch similar to the one I had taken the day before, I had two wonderful dreams—between noon and four o'clock, that is to say, within less than six hours. First, I dreamt that I was an adolescent and was very happy about that. Next, I found myself a member of the household of the greatest king. I had assured him of my patience and unfailing loyalty, and he had by some tacit agreement received me into the circle of youths attending him. I served him with the greatest devotion as he seemed to believe and trust my words. I felt safe and happy because I had been cured of all the infirmities and emotional disturbances from which I had suffered. I was reliving my adolescence, as it were, with its natural vigor and the capacity to patiently tolerate inconveniences. Then one day, the king approached me with an immoral amorous gesture. I did not respond in kind, and after having skillfully extricated myself from the situation I related the incident to a certain old woman whom I knew only in my dream. Thereupon I began to feel great remorse as well as great fear. I thought: if I let the affair be known, the king will have me killed. If I keep it secret, the old woman will no doubt talk about it, and if she should be betrayed I shall be betrayed as well. I, having been so intent on unwavering devotion, shall in the end appear a traitor to my master. If I comply with the young man's wishes, it will be impossible to conceal it. If I refuse him, he will not want to have a witness to his excesses,

since he already committed a misdemeanor with me. I shall perish no matter what I do. My unfaltering loyalty will be despised by everyone as having been infamous and deceitful. Alas, there was no way out of the bad entanglement fate had led me into.

Shortly thereafter I dreamt of being with the said prince—or perhaps it was with his personal physician—I am not sure about that anymore. While in the castle, I was called for by a poor woman, apparently in my capacity as a doctor. If I went to her, the financial reward would be very small. In order to ensure that the woman would not look for me in vain, I left a youth behind as a servant, and a horse or a donkey (I do not recall exactly which). This way I could be certain to find her. Meanwhile I had an argument with my prince on account of this woman. I apologized to him for having been prompted by compassion to attend to her. With this I woke up.

The king's amorous adventure signifies religion, in the name of which we are led into temptation (tentatur). This will cause me considerable worries, but I shall not come to any actual harm. The king himself signifies the practice of medicine: I am to return to the art of healing and thereby be rejuvenated. Thus I will be able to live with a reasonable degree of honor. This is made clear by the second dream, which is really an interpretation of the first. I shall visit only a few patients since I dreamt that I was called upon to exclusively attend a king. There will, however, be crowds of people whom I will not receive. It is furthermore clear that all this is going to happen two years from now: therefore I shall live and be well for the time being.

What is amazing here is that Cardano interprets the king's love affair as a symbol of religion. He obviously adheres to his rule of observing what things are innately and distinguishing this from the meaning they have acquired through laws and customs. He also says that we are tempted in the name of religion, and he uses the word "tentare" here, which can also mean "to touch." This is obviously a premonition of the troubles that were going to beleager him in the name of religion. This did not occur within the following two years, however, but eight years later when he was actually imprisoned. Yet in the end he never came to any real harm. Since this dream was dated 1561, and since it was published along with an interpretation in 1562, Cardano could quite obviously not have had any knowledge of the events of 1570.

The dreams presented above all show that Cardano has indeed regained his composure, and that he can again turn toward the tasks of practical everyday life.

The last example I would like to pass on to you is the "great dream" with which Cardano concludes his book.

6. Finally I am going to relate that extraordinary dream I had on the twenty-third of February 1562. It will give me the opportunity to discuss one of the greatest difficulties of proper dream interpretation: how to distinguish between memory image and dream vision. When these things cannot be separated—and this is frequently the case—my method of interpretation becomes much more difficult and may even look fraudulent. Most dreams contain remembrances of the past; this obscures the whole dream, not only because it interferes with information concerning the future, but also because people realize that all dreams are in part stimulated by things we remember. Since these elements often constitute a substantial part of the dream, people tend to think that the entire dream content is composed of memories.[9] In this dream I went to check how many rings I own. I excluded two rings from the count: one with an amethyst which I always wear, and another one with an elongated red stone which had little value, or so I thought at least in the dream, for I do not own such a ring. I had given this ring to a Spaniard dealing in precious stones, who remained

unknown to me in the dream. I had gone to him merely to have the stone appraised and had not actually given the ring to the dealer; he just claimed that I had done so three days ago and that he had subsequently sold the ring. He furthermore assured me that the conditions of the sale had been advantageous and that I would make a profit. The price was to be paid within two years; two people had jointly vouched for it, although not in writing. The price was set in gold—I do not know how high—but the dealer mentioned the figures forty and thirteen percent and praised it as a most wonderful top price. According to the jeweler, one of the buyers was a silk weaver. He made no further mention of the other who was also unknown to me. I reproached the dealer for having sold the stone to people without means who incur debts. But he assured me that this weaver was a very wealthy man. As he kept on talking about all the gold that would be paid I asked him about the current market value of precious stones. I added that there were certain stones for which I would pay a price exceeding their estimated value. I told him that—among others—I had purchased a moonstone for fifteen gold crowns which he had then appraised at only three crowns. To this he just replied that it was my own business if I wanted to pay more, and that he was content to settle for the estimated value. I now regretted having sold that ring to strangers. Even worse, I had nothing in writing and the buyers lived in a distant country. If they did not keep their word I would have to go all the way to Spain to make my claim. It would mean either to neglect all my other affairs, or to let the matter go altogether. I had then, in effect, for little and uncertain profit thrown away rare stones which it had taken a lifetime to collect. Even a successful conclusion of the sale would mean a substantial loss over what I had paid for them. The dealer's reply to my complaints was simply that I had acted of my own accord, that he had indeed persuaded me to sell the ring, but that he had done so in the belief that it would be in my best interest. I consoled myself thinking that all this had not really happened at all. With this hope I said confidently that I would probably find the ring in the safe, for I had no recollection of ever having gone to the safe and taken the ring out. Yet, that man kept on insisting that the rings were no longer in my possession and that I had in fact given them to him. Then I asked him his first name. He seemed ill at ease and reluctant to speak. Finally he said that he called himself Stephanus. When I asked him his family name he was even more reluctant to answer, but at last he said: de Mes. (Mes[10] is the name of a town the emperor had laid heavy siege on a few years ago, with the king of France defending it successfully.) Finally I asked him if he was going to stay in Milan. He said that he did not have a permanent residence and was traveling about. Then he said that he was going to live in a nunnery. He also mentioned the name but I do not recall it. He added that I would always find him in Jewellers' Row. I replied that in this case I was willing to trust him. Besides, I was consoled by the thought that the jewels were really still in my possession.

Meanwhile I had the impression that there were several large beds on the upper floor of the house. I also believed there to be a chained baby elephant and another very beautiful animal with a spotted coat like a panther's. I do not recall all the details nor the name of the animal. Yet it was on account of these animals that I took extraordinary precautions. I tamed the elephant by offering him some bread, which gave me great pleasure. Both animals were looked after by Hieronyma, an old maidservant who no longer lives with me, but rather somewhere east of here. We had talked about precious stones the previous day, and the dream could therefore be characterized as a typical dream with a memory stimulus. But there are also elements referring to the future. It contains a warning not to get involved in unsound undertakings which—although they may for the moment give reason for hope—nonetheless present various hazards and risks as well, particularly concerning the completion of my books: they are the jewels and the money I was worried about. Moreover, the dream makes it clear that it is entirely up to me to decide on the right course of action, because I was hoping all the time that no business deal had taken place and that the stones were in my possession. The dream obviously admonished me to make decisions according to my own judgment. The meaning of the two animals is clarified by the presence of the maidservant. Since they are both young animals they represent youths. The elephant is my son (the younger one who stayed alive); the second animal represents a reader who, by virtue of good taste and diverse knowledge as well as his special musical abilities, is proving to be a most excellent and versatile companion. The rest of the message can be deduced from this. Names and first names mentioned are ruled by the center. I am told that I must serve a prince or else I shall lose all my possessions.

This dream is related in a most vivid manner. The element of uncertainty—the question of whether Cardano still owns the stones or whether he has indeed sold them—is communicated admirably. In the "dream encyclopedia" precious stones are in fact listed as symbols of books. This is not surprising, since books were extraordinarily costly at that time; besides, Cardano placed the greatest importance on his own literary activities, which he hoped would immortalize his name. It is particularly noteworthy that he deduces from the elements of uncertainty prevailing in the dream vision that his future situation will depend entirely on his own actions. The decision he faces apparently concerns the professorship at Bologna. He did go to Bologna, as we know, despite the warning signals given by his dream—and his sojourn there was indeed fraught with a variety of perils. The elephant stands for his younger son, a real problem child who turned out to be a good-for-nothing. On several occasions Cardano had him put behind bars and he finally disinherited him. The other youth—symbolized by the panther—is one of the young people whom Cardano took into his house. He instructed them and they in turn served as secretaries and readers. In chapter thirty-five of *De Vita propria* some of these youths are mentioned by name.

The statement, "names and first names are ruled by the center," does seem rather obscure. Well, the jeweller calls himself Stephanus de Mes, and as the parentheses indicate, Mes is the town of Metz, of which the protomartyr St. Stephanus was the patron saint. In Latin the town is called *Medio*matricum, and the jeweller is staying in Milan, *Medio*lanum in Latin. So the names and surnames point decidedly to a medium, that is a center. This "center" was the place of contest between the emperor and the French King, an event of considerable historical significance. So Cardano concludes: names and surnames point to a center—are ruled by a center—which has to do with great princes, the emperor and the king. This means that he has to serve a prince, which is of central importance. Cardano, writing for his contemporaries, seems to think that his cryptic intimation would be understood by his readers.

I hope that my selection from the large number of dreams Cardano relates and analyzes conveys something of this side of his personality. Some dreams, like the third example, are short and easy to understand. The narrative is always very graphic. Obviously, Cardano has observed everything most keenly, his mind being alert even during sleep. He also makes it a point to state whenever his recollection is not quite clear. It irritates him, for instance, that he cannot remember whether he had been riding a horse or a mule, or that he has forgotten the name of a convent that was mentioned in the dream. Cardano strives for scientific precision in his accounts and he sounds absolutely credible. The dreams project coherent experiences of great plasticity. The dream events are often quite dramatic, and Cardano is reflecting on them as they occur. He has cultivated an expressive and figurative dream language suitable to present the dream content most accurately. At the same time, the reader becomes more intimately acquainted with the author as a private person. We learn details about his life and habits: what he eats, that he likes to take long naps after lunch, that he is a collector of precious stones because he delights in them. We can follow his reactions, his thoughts, and how he is overcome by intense emotions.

The dreams are full of fantasy but in no way fantastic. They deal with real-life problems, and Cardano interprets them accordingly. Although Cardano also shows certain dream elements to be obvious references to the future, his dreams and his analysis of them primarily help him to endure life, to take new courage or to resign himself to things as they are; in other words, they are an aid in coming to terms with himself and the world at large. As the future is in many ways determined by the present, Cardano can justifiably say that dreams point to the future. The main objective of modern dream analysis—regardless of the school of psychology involved—still seems to be to help people to come to terms with their inner and outer world, to find ways and means to live with themselves and the world, though it may be impossible to effect any change.

Our examples show that for Cardano, reflecting upon his dreams was indeed therapeutic. It helped him to deal with the tragedy of his son's fate; beyond that, it helped him to assume a clear and constructive philosophical stance. This is well-illustrated, for instance, by the developments in the fourth and the fifth dream (the attack by the senator and the king's amorous approach). Cardano understood the meaning of the first dream sufficiently well to have the second dream the following day. A person who cannot free himself from his neuroses will not dream in this way. These dreams show a man with a remarkably fine relation to his unconscious, but there is absolutely nothing about him that one might call pathological. In his own time, Cardano would probably have been characterized as a person of melancholic disposition prone to choleric outbursts. Such a man was regarded as the prototype of the scholar. Cardano's admiration for the graphic art of Dürer brings to mind that artist's "Melancholia" (Melancholy), or his "Nemesis", also called "The Great Fortuna." Cardano's dream imagery shows a distinct kinship with the figurative elements of these works of art.

7 *On the Art of Living with Oneself*

AS I MENTIONED earlier, Cardano spent the last years of his life in Rome. He was given a pension by the Pope and was permitted to practice medicine, but forbidden to publish.[1] No further legal proceedings were begun against him, and the matter was quietly laid to rest. However, Cardano was still worried that the Inquisition might prosecute him again. The new pope, Gregory XIII, was regarded as amiable and had a liberal reputation, but he marked the beginning of his pontificat in Paris on St. Bartholomew's night 1572, celebrating the festival with a TeDeum. The intellectual climate had changed significantly during the previous decade as the counterreformation was spreading everywhere. Understandably, these developments were most disquieting to the old Cardano, and he was not always successful in maintaining his philosophical calm. He was, nonetheless, a true philosopher and experienced psychologist, and knew how to cope with his own troubled mind.

This is well-illustrated by a strange dialogue[2] between the author and his father who had died long ago, which Cardano transcribed in his seventy-fourth year. The dialogue can be viewed as a philosophical—contemplative text similar to other essays of his—such as "On Consolation," "On Human Counsels," or "On Death." The dialogue is a good example of this side of Cardano's literary activity. It is not, however, a literary text in the usual sense, as the author addresses himself rather than an audience, so that the dialogue is essentially a soliloquy which—unlike the platonic dialogue—is not intended for bystanders.

Cardano begins by lamenting his own precarious situation in Rome, which he finds utterly depressing. Then he sees an apparition of his father's spirit as it had appeared to him some forty years earlier. This recollection is the starting point for his meditation. The father embodies the prudence Cardano has apparently lost in his state of depression, and which he now confronts with a kind of obstinate helplessness. In the course of the dialogue it becomes clear that at the root of his depression, in which he only perceives his present miserable state, lie unconscious fears concerning the future. In the final analysis it is fear of death that has taken hold of him—not, of course, unusual for a man of his age. The father now proceeds to point out that death also means rebirth and metamorphosis, and that this process has now begun for Cardano; therefore, his fears—when interpreted correctly—are fears of parturition. As regards death, the father admonishes him that any sensible man ought to be liberated from all his sorrows at the thought of his own death, as death is the guarantor of freedom. Cardano should, therefore, recognize his lamentations and fears as childish and absurd, and while he lives he should devote himself to practical tasks, particularly the care of his grandchildren. Thus, the dialogue leads from fear and anxiety by way of contemplating death and rebirth back to life.

The way in which Cardano deals here with depression and anxiety might be called "active imagination" in the Jungian sense.[3] In this meditation, an autonomous and superior function is accorded to the unconscious, symbolized by the father-image, in relation to ego-consciousness. Accordingly, the imagination personalizes the unconscious, thus allowing for an actual contact and dialogue. In its course the meditator confronts his own weakness, but at the same time he becomes aware of a new strength, although not his own, for—as the father says—goodness and justice come to us only by the grace of God; to Him we owe far more than to all our own abilities.

I am going to render the dialogue in as literal a translation as possible[4], and I hope that the reader will neither mistake it for a piece of literature, nor for pure philosophical reflection. Rather, it is a testimonial on coming to terms with one's own existence.

Dialogue between Girolamo Cardano and His Father, Facius

G: Alas, what an unfortunate situation I am in. Through injustice I have been deprived of my worthwhile and honorable teaching post and at the same time I have lost the opportunity to publish my books. Worse still, at an age when others retire, I am forced to tolerate these Roman doctors; already Galen made it clear that he could not tolerate their betters. I have no income, I am separated from my children and grandchildren, I am in poor health, and neither the loyalty of servants nor that of friends can be relied upon. Like that blind man I can say: "Quis errantem Oedipum parvis excipiet donis?" But even he was not nearly as badly off as I am, for he had many descendants, his sons were kings, and his troubles were not the result of an insulting injustice. What happened to him was neither an accident nor the fault of others. He brought it upon himself and hence had to bear it. Is there any man more unfortunate than me? What am I to do?

F: Who is that lamenting there so loud and foolishly?

G: Woe is me, for now I am being pursued even by demons!

F: I am not a demon, I am your friend, your very best friend. I used to be your father but have long since cast off this sordid mortal frame as you know: it is I, Facius Cardano.

G: Be blessed, father, and have pity on your only son; help me, if you can, you have come at exactly the right moment.

F: You are not really as unhappy as you claim to be. Rather, you are making yourself miserable.

G: I did not mention just now what I heard in a dream[5]—I felt abandoned at the time—and you appeared to me as you had done on previous occasions. Your message then seemed to be that I could neither stay here nor leave, as neither course of action would be safe, honorable, or proper.

F: I am glad that you recall that dream! For if you heed the signs you will know that the time has come when you have entered the heaven of the moon. As you have just read in Beatus[6], the spirit of the moon holds the lowest place among those substances which are completely incorporeal. Although he can perceive himself by himself, he can do so only imperfectly, and not as clearly as the higher intelligences. Our own intellect is still further removed from the "First" than that of the moon. For that reason he cannot perceive himself by himself. He can only perceive himself by perceiving other things. We cannot perceive other intelligences because we cannot perceive ourselves intellectually, for our intellect is bound to matter, and anything that is completely separated from matter remains inaccessible to him. Of such things we can gain knowledge only by analogy, by interpreting similarities, and this only in a most general way. But the spirit of the moon perceives himself by himself, and since he is connected with the "First," he perceives others as well.

All this you read in Beatus, did you not?

G: Indeed!

F: Today is the fourth of April 1574, and you shall regard this day as your first, so to speak, new birthday. You understand now how you will—starting from the sphere of the moon—ascend through sphere after sphere until you reach the highest one. Rejoice that your dream is becoming reality and that the time is now.

G: What shall I be happy about?

F: About the knowledge that God has endowed you with a special gift. What hardship or acrimony could you have suffered, what might still happen to you that you would not voluntarily, even gladly, take upon yourself? Can you not see how God is taking care of you? This is indicated, among many other things, by that dream[7] which you think so full of forboding, so gloomy and frightening. Further evidence is provided by the letters that fell to the ground, one of which moved upright towards the table leg. And while you were writing another letter, the letter sander was concealed by it, which is ordinarily, of course, impossible as one cannot write in this manner.[8] Finally, it is also indicated by that clap of thunder which threatened the house for so long, and by that frightful uproar at the end of the third day when you had been incarcerated. It sounded first at the prison gate, then at the window through which sunlight still fell. The window-bars were so shaken that they clattered. The physician Rudolf Sylvester was a witness to these wondrous occurrences.[9]

G: All this is true, but I fail to see its use. And yet I am consoled by these things, and with the poet I say: "Tyrant, your ferocious speech does not frighten me. It is the Gods I fear, the wrath of Zeus."

There is something which distresses me far more than

the loss of my professorship and the fact that I cannot publish my books. It is this: on the day and at the place of my arrest, but at a different hour, I encountered two single women, tradeswomen most likely, who were creating a mighty uproar with ropes and dogs. I was ignorant of my fate then, but this scene was probably meant to astound or frighten me. The blows struck by the second one resounded most disagreeably in my direction. Subsequently, there were other omens, which I have just mentioned. These omens confuse me so much that I am in constant fear of being arrested again and that in the end I shall be publicly executed. This causes me great anxiety. I do not have a sense of guilt, there can be no question of that. Neither am I afraid of any earthly authority—although I know it is the highest one.[10] Nevertheless, this matter is occupying my mind because I cannot explain it to myself. I am also disheartened to see that my adversaries consider what has happened to me as a result of my piety to be the result of either my cowardice or their chicanery.

But to get back to the main point: I am distressed that I have been unable to find an explanation for these miraculous signs, and I have never found anybody who has been able to explain the matter to me. I told the story to many to see if anyone could provide an explanation and thereby free me of my anxieties. Yet one thing is apparent: telling all this does not enhance my reputation; people think me either boastful and fanciful, or else silly and superstitious. But I have to disregard this, because in view of my fear of imminent disaster, the well-being of my family as well as myself are of foremost concern to me. If you have some advice, father, please tell your son: is it ordained that still more and greater things are going to happen to me?

F: I shall be glad to tell you. But first I have to admonish you on two accounts.

The first is that people who do not believe that there are greater things in store for us will think non-essential things important. Those who think that the hereafter is a nothingness will take mere vanities as true being. But you know better, you know that it is the supreme thought, and therefore you also know that this life is but a dream and a shadow. This becomes perfectly clear to anyone who reflects on the past. Therefore, why do you torment yourself like that, like someone who envies a boy who has cherry pits to play with while you have none? What can you gain in this life that is lasting and significant, especially if you compare it to that other life? How could you get yourself into such an unfortunate circle where you err and grieve for no reason and do not find any happiness? What is there in this life that is beautiful and permanent and safe from the corrosion of time? Is there anything that improves with the advance of time? Wealth increases itself, but when you are old others will reap the benefit from it. As you can no longer enjoy it, it will only sadden you. Let me therefore ask you first: what is there that you can rightfully complain about?

The second is that I ought to explain to you at once the entire course of history and point out the purpose of each single thing to show you the criteria by which the spirits are judged. Without this, nothing truthful or conclusive can be said about what the future holds and about the proper course of action. We would not understand each other and help would avail nothing.

Let us consider first worldly possessions: think of what has been taken away from innocent people in wartime, in the past and in the present. It is happening in the towns of Hungary and Cyprus today. You have lost nothing, although you are not altogether innocent, while those innocent people have lost all!

G: I am not complaining of having lost things, but of the fact that I have nothing left and that I no longer have the chance to recoup my losses.

F: If your sons are worthy they will be content with what you have. If they are incompetent, all the things you might still acquire would not be enough for them.

G: Then this shall no longer worry me! But I am also concerned about my reputation: I have been forbidden to publish my books. I shall see people inferior to me in teaching positions.

F: But you know that you must die! And you cannot be putting your hopes on uneducated or even on educated men who will become professors.

G: Indeed not!

F: Be happy instead that you have lived this long. You have certainly reached the age to which scholars usually live. Solon has already said that life lasts seventy years. This is the age at which he died; Socrates also died in his seventieth year. Only Hippocrates is said to have lived longer, but he did so only to endure further misery, he is supposed to have died completely senile. The others all died at an earlier age: Aristotle, Galen, Avicenna, Phericides, Pythagoras, Virgil, and Horace. "Immodicis brevis est aetas, et rara senectus".

Besides, you did not read a longer life for yourself in the stars, the lines of the palm did not indicate it, nor did physiognomy or the way of life. As regards the wrongs you have suffered: if they happened by chance it must not distress you, for you know that this comes from God. But if you are being hurt by mere mortals, cannot God manipulate them, just as he does an illness which snatches one away or renders one helpless?

The Cyreneans said that pleasure and pain are felt both in the body and in the soul. This is true, but pleasure is felt more in the body, and pain more in the soul, for the latter is rather illimitable and sees all too many things in direct relation to itself. To all experiences it adds its own apprehensions and memories, thus increasing grief immeasurably. Although the individual soul testifies to the brevity of life, it imagines that the life, deeds, and resolutions of others are forever lasting. It will remember injustice, insults, and distrust, but it will forget kindness, benefaction, and tokens of love; it will make a mountain out of a molehill.

All this is not serious, though. After all, the most important thing is that you are alive and well. Despite everything you would rather be alive than dead. Grasp the fact that life has always been like this. Wanting to live without being prepared to suffer is asking the impossible.

And once more I want to remind you of the resurrection. Tell yourself that you have already lived three and a half years beyond the normal life expectancy. The past is assured, and the future cannot hold anything to drive you to despair. You have three descendants, many powerful friends, a great name, and great knowledge. You have published 131 books, and another 111 are yet to be published. You have knowledge of many things which are concealed to others. Your inheritance has quadrupled. What the ignorant will think a misfortune—that you are seventy years old—you must consider as your greatest asset. For all hardships are behind you now, the good experiences are yours forever and will remain with you as gratifying memories. Yet the ultimate faith you must keep should remain the Adrastia[11], the unchangeable order, the eye that sees all. Kingdoms are overthrown more easily than wisdom is deceived. Love does not aim for the best, or else the possibility is wanting to answer just pleas.[12] Three things have yet to be discussed—for vengeance is God's—: the increase of the inheritance, the professorship, and the publication of books.

As far as the inheritance is concerned, it is more important for the children to learn to make proper use of it than for you to use it. It is more important for them to want to keep it than for you to increase it. As regards the fame you can gain from your books, let me quote Zeno:

"Just as I was sailing with favorable winds, I got shipwrecked. We would all have had to perish had not some of those among us been pious."

This is how imperfect these books were: they contained basic flaws and errors. And what is to be said of the sons? Did you not lose them because of carelessness and indulgence? And are they not now in safe keeping!

G: It is constant worry that wears people out. I could suddenly be struck by disaster or suffer a stroke of bad luck.

F: What disastrous thing could possibly befall you? You have never been ill, you still read without glasses, neither your bladder nor your mind have weakened, you have no trouble letting water. You do not suffer from insomnia, and there is no hard swelling in your abdomen.

G: I fear that all this has just been postponed by a temperate life. But there are no permanent rewards in life.

F: There have been famous people who spent their lives grieving, but nevertheless lived to an advanced age. There are also women among them, Pomponia Graecina, for example, who for forty years mourned the death of Julia, daughter of Drusus. She would only hear mournful speech and tolerate only drab clothes. The most serious charges were brought against this woman, yet she lived well into old age.[13] If one scorns sensual pleasures, one should do so not from sorrow but from the understanding that they are shameful. Besides, any sensible person should be diverted from a sense of grief by the thought of his own death. Defiance of death will give him strength. For death is the guarantor of freedom.

Now that you have understood this, I can proceed to the interpretation of your dreams and presages. That earlier dream—you came to alter a seal, a pharmacy was along the way,—what was the significance of the man with the cow? He may be regarded as accidental; at any rate, he has nothing to do with your arrest. He might signify death, but only the death of vice. All this should be looked upon as an ancient ritual of penitence, as a fairy tale. The other matter is of greater significance. Those knocks at the entrance gate signify a verdict in Bologna, those at the window one in Rome. The clatter of bars means that you had to depart from Bologna, your new home,[14] in order to enter on a glorious pilgrimage. But there is no danger of a second arrest, because the portent occurred not at the beginning but at the end of your imprisonment.

There still remains much in which you may rejoice. You are distressed by the false assumption that once there lived people who were really good; but you are wrong, they were all the same: Aristides, Socrates, Plato, Cato minor. And yet, besides the many motives that affect the intellect, such as anger, envy, ambition, avarice, and vehement passions, these men did have an incentive for goodness. The rest of mankind simply acts out of ignorance and the lack of having experienced goodness. Only by the grace of God can man truly be just and righteous. To him we owe far more than to our individual talents.

Since the basic requirements for your ascent toward the supreme state have already been fulfilled, you must now first take care of your financial affairs, and then of the education of your grandchildren. From this, everything else will take its due course. You will yet experience great joy.

G: I thank you, father. You have shown me that despite deep distress there are still many comforts for me: the thought of my own death, the image of the resurrection, the thought of a transition into the totally other, the laying down and crucifying of the body.

F: Farewell, then!

G: You are leaving so soon?

F: I shall come to you again.

G: I am very glad of it. But where is my son?

F: He is here with us. All is well. And you—farewell!

Postscript

WHEN GALILEO set out to found a new science, he firmly denounced all forms of occultism. Physics was to become a mathematical science. Consequently, Galileo accepted only dimension and motion as primary qualities of a body; these he considered mathematical qualities. God is seen as the great geometer who has given measure to all things. Galileo did not presume to say that all aspects of the order of the universe could be explained and understood on this basis. The abundance and diversity of phenomena seemed to him beyond the capacity of human mathematical thought. But since there are mathematical laws underlying all phenomena, it seemed reasonable to select individual and isolated phenomena and to represent them mathematically, much the same way Archimedes did. Galileo believed that this would at least provide complete and definitive answers to specific questions. The basis of Galilean physics was the idea of "natural" motion. It is an Aristotelean concept, but Galileo gave it an important new meaning. His cosmology is founded on the Copernican world system which designates the circular orbits of the planets around the sun as "natural" motion requiring no further cause. Now the earth is also a celestial body. Therefore, if the earth was a perfect sphere with a perfectly smooth surface, then a body on it, once set in motion, would move with a constant velocity around it. This movement is circular and "natural" and corresponds to the "law of inertia" as Galileo understood it. He believed that by applying this law he had found a theory of tides of which he was very proud.

He asserted that this theory proved the earth's movement around the sun as well as its daily revolution. To Galileo, the notion that tides are caused by the moon, which as a moist celestial body affects the volume of the earth's bodies of water, belonged in the realm of medieval superstition. He knew, of course, that the tides are determined by the lunar cycle rather than by the course of the sun, and that they vary in intensity depending on the phases of the moon. He thought this was caused by the rotation of the earth and the moon around their common center of gravity. He conceded that this did interfere to a degree with the basic effect he was trying to explain through his theory, but he argued that it did not significantly alter it. Galileo could not understand that Kepler shared the traditional assumption that the tides were the result of the moon's effect on the oceans. But his own theory of tides is wrong—being based on an incorrect application of the law of inertia he himself discovered. The traditional assumption that the moon determines the tides, which Kepler shared, is, of course, correct. By purging science of all aspects of occultism and by attempting to explain everything purely on the basis of mathematical principles, Galileo did—as Kepler would have put it—throw the baby out with the bathwater.

Contrary to Galileo, Kepler accepted the natural philosophy of the Renaissance[1], and he believed, as Cardano did, in astrology.[2] However, in order to explain the influence of the stars within the framework of the Copernican system, Kepler had to develop a special theory.

In 1602, he published *De Fundamentis Astrologiae certioribus*. In it he states: there can be no doubt about the influence of the stars since the sun's position causes the seasons. Without an outside source of heat, the air, earth, and water will of their own grow cold. From this we may conclude that matter is naturally cold. For Kepler, as for Cardano, cold and dry are essentially privations, and are therefore not to be attributed to the celestial bodies. Kepler also believes in the effect of the moon on the oceans: moist bodies expand during waxing moon and contract when the moon is waning—experimentia probatum est.

Kepler replaces Aristotle's theory of the four elements, which he considers to be wrong, with another one, founded on Platonic principles: Idem et aliud. The Alteritas is identical with matter, which in turn contains—so to speak—geometry.

The effect of the planets is not, however, a formative process, but rather, a stimulus for moistening and heating matter. This is brought about by light which is warm and which—when reflected—produces moisture. These ideas are in perfect accordance with Cardano's philosophy of nature.

In 1604, Kepler published *Ad Vitellionem Paralipomena*. These are supplements to the classic medieval textbook on optics by Witelo. Kepler departs significantly from the original presentation insofar as he prefaces the part dealing with physics with a lengthy introduction reflecting the ideas of the sixteenth century, rather than those of the Middle Ages. It is true that Witelo also begins by explaining what light is and praising its almost divine qualities. But compared with Kepler, Witelo sounds extraordinarily matter-of-fact. This is the way Kepler puts it: light was created by God in a process reflecting the Trinity. Light aids in producing perfect relationships among the bodies. These are enclosed within their surface area and therefore cannot multiply themselves (multiplicare in orbem).[3] Yet they contain forces within them that are freer and not of bodily substance. Their substance is spatial, and it affects the surrounding area as it streams forth. This phenomenon can be most easily observed in magnetism. The highest principle of this kind is light, which gives life and form to all things; it is the bond between the physical and the spiritual world. Light expands on a spherical surface, and for this reason its intensity, that is to say, the density of the light rays, diminishes with the square of the distance. Therefore, light is a "surface-effect," and its intensity is a surface-intensity. And because only things of the same kind act upon each other, the surface of the bodies acts upon light. Although Kepler states here in effect the fundamental law of photometry, the introduction as a whole does not deal with physics in the modern sense. The edition of the *Ergänzungen zu Witelo* in the series "Oswalds Klassiker" actually omits the entire introduction! Many physicists find these "occult" elements in Kepler's thinking objectionable indeed. For Kepler himself they are, however, a source of animation. One of his great contributions to astronomy is his second law: the radii from the sun to the planets sweep over equal areas in equal time. Kepler did not discover this through observation—this was impossible anyway because the distances of the planets from the sun were not known accurately enough.[4] Kepler saw in this principle a fundamental law of mechanics: the sun propels each planet by a magnetic effect, as it were, which intensifies with the planet's proximity to the sun. It is a case of action at distance—in other words, an "occult phenomenon." Unlike Galileo, Kepler was not in the least perturbed by this. While Galileo in his uncompromising rejection of occultism held fast to the notion that the planets' orbits are circular—which is natural if the planets move "of themselves"—Kepler was led by his law of areas to the discovery that the orbits are elliptical. With this he overthrew the long-standing platonic prejudice that only the perfection of circular motion was appropriate for the heavenly bodies.[5] Kepler, unlike Galileo, believed that bodies are "by their nature" inert.[6] From this concept of inertia, which Galileo did not refer to at all and which Descartes rejected, Newton developed his concept of the inertia of mass that causes bodies to resist changes in velocity. Newton calls this inertia the "vis insita," the "inherent force," a designation that still clearly reflects the occult origin of the concept. Inertia and long-range effects are physical concepts which cannot be regarded as pertaining to "primary qualities" in the Galilean sense. They therefore also lack the "clarity" and "distinctness" that Descartes demanded of fundamental scientific concepts. In the final analysis, Descartes accepted only geometrical concepts—extension, impenetrability, and motion—as meeting these criteria. Thus the rudiments of Renaissance philosophy were radically simplified by Descartes: there is only matter and extension on the one hand, and the mind or power of the intellect on the other. Both these principles are Platonic legacies, and we encounter them in Cardano as well as in Kepler, except that Descartes has divested them of their "tools"—namely, moisture and heat. Consequently, his doctrine lacks a dynamic principle which causes and directs all occurrences. For this reason matter has to constantly keep up the motion instilled by God. In variance with Galileo, Descartes believed that he could design a comprehensive cosmology based on his principles. In this he proved to be more antiquated than the Florentine physicist. But Descartes's project was doomed to failure from the beginning, since his principles are inadaquate even with regard to Galilean physics, which limits itself expressly to finding solutions to individual problems. Kepler's ideas, in the final form Newton gave them, made possible the development of mechanics as an adequate tool of physics. As a stimulus to the imagination, occultism gave Kepler's thinking its particular direction. It is not surprising then, that his mentality is seen as peculiarly ambiguous.

Newton's contemporaries were still familiar with the source of his ideas, and this was often the very reason for them being so passionately rejected. Newton's concept of force[7], and his theory of gravitation in particular, seemed to many unauthenticated. Christian Huygens was one of the critics, but he was far too good a physicist not to appreciate the importance of Newtonian mechanics. He could not, however, accept the idea that every particle attracts every other, "since this could in no way be explained by the principles of mechanics."[8] In a letter to Leibniz dated November 18, 1690, he calls Newton's theory of tides totally unsatisfactory, like everything else based upon Newton's principle of attraction, which Huygens thought absurd.[9]

Leibniz himself attacked all of Newtonian physics most vehemently. There were also personal reasons for his opposition, but it is noteworthy that he reproaches Newton above all for a relapse into occultism. Newtonian mechanics is indeed contradictory to a Galilean or Cartesian system, and his concept of force is obviously related to the occult forces considered by the natural philosophers of the Renaissance. In his *Antibarbarus Physicus pro Philosophia Reali contra renovationes qualitatum scholasticarum et intelligentiarum chimaericarum*[10] Leibniz gives a well-aimed and spirited presentation of his views. He says:

An evil fate wills it that men will from time to time revert to darkness out of boredom with light. Ours is such a time, with great opportunities to learn the right things being spurned, and a wealth of the most lucid truths being disregarded in favor of obscure trivialities. People are so eager for change that they are prepared to turn their attention away from a profusion and diversity of fruits and waste it on acorns. A physics that explains all phenomena of the physical world in terms of number, dimension, and weight, or in terms of size, form, and motion, which furthermore teaches that all motion is caused by momentum, thereby tracing all physics back to mechanics—thus making it comprehensible—is too clear and simple for these people. Instead they revert to fanciful ideas.

After some polemics against those physiologists who do not regard an animal as simply a "divine mechanism," he continues:

There are those who prefer to turn their attention back to qualities of the occult and the prevalent disciplines of scholasticism. But because these were subsequently discredited by barbaric philosophers and physicians, they are today given new labels and referred to as "forces." In actuality, there is but a single kind of physical force, namely that produced by momentum. Yet, these people invent special forces whose properties they change when expedient. They speak of attracting and repelling, adjusting, expanding and contracting forces. Men like Gilbert, Cabaeus, or Honoratus Fabrius may be forgiven for harboring such notions, since philosophical clarity was not a mark of their time, or at any rate not yet sufficiently appreciated. But how can any reasonable person today subscribe to a belief in fantastic qualities that is tantamount to a betrayal of all natural principles? One can, of course, admit the existence of magnetic, elastic, and other such forces, provided they are not regarded as original and given phenomena but are understood to be the result of motion and form. But the patrons of these fashionable new ideas are unwilling to accept this condition. Notions of the planets gravitating and striving toward each other have now been confirmed as correct. But this has given rise to the fiction that matter is provided by God with a power of attraction, with some kind of sympathetic force which functions as a power of perception or a form of intelligence informing each particle. By virtue of this special power, matter is supposedly able to perceive and covet even things which are remote. Such beliefs do not, of course, allow for basic mechanical considerations. These would show that the mutual attraction of large bodies can be explained quite simply by the motion of a fine body penetrating them. The patrons of the fashionable ideas enunciate still other occult qualities, and all this is bound to lead us completely into the realm of the obscure. . . .

Leibniz finally exclaims:

What would Descartes and Boyle say if they were to return to find those fantastic notions in vogue again which they refuted long ago?

Although it is perfectly correct to say that gravitation as well as the electrical and magnetic forces of the new physics can be called "occult" properties, as they cannot be explained in terms of simple mechanics, it is hardly justified to call it a throwback to medieval obscurantism, when Newton and his followers did not endeavor to complete the Cartesian program. After all, Descartes's doctrine is not nearly as simple and clear as Leibniz thought. There is no reason why the infinite variety of phenomena should be explicable within a framework of concepts which seem "natural" according to one particular established philosophical system. To someone like Cardano, the idea that all phenomena of the physical world can be explained in terms of extension, form, and motion would have seemed quite incredible, and we do not believe this today. Incidentally, Leibniz in particular thought very highly of Cardano. In the *Theodicy* he says:

> It seems to me that knowledge exerts a magic that is comprehensible only to those who have been seized by it. I am not only referring to factual knowledge which is not concerned with the powers of reasoning, but rather that kind of knowledge which Cardano possessed. He was a truly great man, despite his numerous misconceptions. If not for them, he would have been unequaled.

It is indeed striking how closely Leibniz's monadology resembles Cardano's conception of souls. Central to both theories is the idea that the physical world is at once unity and multiplicity. In Leibniz's system the monads form an All-unity in accordance with the doctrine of the "pre-established harmony," which constitutes a principle of order. Whereas in mathematical theory the "continuum" (chain of continuity) contains the whole in actuality but the parts only potentially, both Cardano and Leibniz assert that in reality the parts of the continuum precede the whole and actually exist.[11] It is to be remembered in this connection that to Leibniz the true continuum—that is to say space—does not have a reality of its own; rather, space is represented by the order of the monads. These are beings akin to souls, and space is the product of their representation. In this respect space in Leibniz is analogous to the *Aevum* in Cardano which is created by the representation of the "intelligences." In view of this, Leibniz's praise of Cardano is hardly surprising.

But what were Cardano's "misconceptions" that Leibniz alludes to? The reference apparently concerns Cardano's occultism, his belief in the sympathetic effect of substances upon each other regardless of physical proximity, as well as his attention to dreams and presages. In the seventeenth century, all this was regarded as superstition, and the prevailing tendency was to free oneself from such notions. But perhaps the baby was once again thrown out with the bathwater in this process of liberation. For we know today that it has nothing whatever to do with superstition if we pay attention to our dreams, although we no longer believe that our significant dreams are affected by heavenly bodies. Instead, we recognize our unconscious as their source. With Cardano we find the unconscious still projected onto the stars, which allowed a unitary view of world. When this projection was discredited, the unitary image of the world disintegrated. Descartes's attempt to provide a theory of the world based solely on considerations of mechanics is ultimately more fantastic and absurd than any theory advanced in the Middle Ages or the Renaissance, because Descartes presents ideas of a rather fantastic nature while asserting that they follow with compelling logic from "clear and distinct" principles.

Since then, no one has succeeded in forming a "scientific view of the world," of the whole world comprising the material and spiritual realm. Anyone who still believes that this is possible not only overestimates the cognitive power of the exact sciences, but is adhering to an archaic delusion: namely, the Cartesian belief in the omnipotence of the intellect. For this reason we should not commit ourselves so exclusively to a strictly scientific point of view that we lose the sense of the immeasurable richness of the world—the spiritual as well as the physical. There are many more things between heaven and earth than the scientific mind can perceive.

Notes

Notes to Introduction

1. Thaddae R. Rixner and Thaddae Siber, *Leben und Lehrmeinungen berühmter Physiker am Ende des 16. und am Anfang des 17. Jahrhunderts* (Sulzbach, 1820), Book 2: "Cardano."

Kurd Lasswitz, *Geschichte des Atomismus* (Hamburg, 1890), Vol. I, p. 308 ff.

2. *Opera Omnia Hieronymi Cardani, Mediolanensis,* 10 volumes in double column folio (Lyons, 1663).

Karl Spon, publisher of the *Opera,* descended from a Calvinist family of Ulm. His father had acquired the rights of a citizen of Zurich. He then went on to Lyons to attend to business affairs. Karl was born there in 1609. In 1620, he was sent to Ulm to attend the Gymnasium, and later went on to study medicine at Paris and Montpellier, where in 1633 he took his doctorate.

In 1635, he became a member of the College of Physicians at Lyons and was given the title of "royal physician." He remained a Calvinist and retained his citizenship of Zurich. He later bequeathed it to his son. As regards his own philosophy, he was a follower of the school of Gassendi. Well versed in several languages—Greek, Latin, German, and French—he published translations of verse, among them works by Hippocrates. He edited most of the medical works published at

Lyons at that time. Spon's edition of the *Opera* was still listed in Wilhelm Heinsius' *Allgemeines Bücherlexikon* (Leipzig, 1793) at the price of 30 Thaler.

In 1967, Johnson Reprint Corporation (New York, London) republished the *Opera,* with an informative introduction and bibliography by August Buck.

3. Galileo Galilei, *Dialogo sopra i due Massimi Sistemi del Mondo* (Florence, 1632), at the end of the fourth day.

4. A. C. Crombie, *From Augustin to Galileo* (London, 1959).
Marshall Clagett, *The Science of Mechanics in the Middle Ages* (Madison, 1959).

5. Bertrand Russell, *A History of Western Philosophy* (New York, 1945), p. 529: "Kepler is one of the most notable examples of what can be achieved by patience without much in the way of genius."

It comes as no surprise, then, that Russell misunderstood and judged unfairly the important book by E. A. Burtt, *The Metaphysical Foundations of Modern Science* (1924).

6. Lynn Thorndyke, *History of Magic and Experimental Science,* volume IV, p. 614.

7. Ernst Cassirer, *Individuum und Kosmos in der Philosophie der Renaissance* (Leipzig, 1927), pp. 159, 160.

8. See Festugière, *La Révélation d'Hermes Trismégiste,* 3rd edition (Paris, 1950), Volume 1, p. 189 ff., and p. 355 ff.

9. A good description of his life can be found in the German edition of *De Incertitudine et Vanitate Scientiarum.* Agrippa v. Nettesheim, *Die Eitelkeit u.s.w.,* edited by Fritz Mauthner (Munich, 1913).

10. He shares this idiosyncracy with Julius Caesar Scaliger (1484–1558), who gained a reputation by his attacks on Erasmus.

11. "I cannot follow Mauthner's interpretation (l.c.) of the work as a "book of confession."

12. Girolamo Cardano, *Podagrae encomium* (Basel, 1566). In 1522, Willibald Pirkheimer wrote a similar commendation.

13. Both the historical Faust and Nostradamus, in whose "personal book" Goethe's Faust recognizes this sign, belong to the generation of Agrippa and Cardano.

14. Lo, single things inwoven, made to blend,
to work in oneness with the whole and live
Members one of another, while ascend
Celestial powers, who ever take and give
Vessels of gold on heaven's living stair,
Their pinions fragrant with the bliss they bear,
Pervading all that heaven and earth agree,
Transfixing all the world with harmony.
O endless pageant!

[Translation by Philip Wayne, Penguin Classics, L.93 (Harmondsworth: Penguin Books, 1959).]

15. *Opera V,* p. 491.

16. Jacob Burckhardt, *Die Kultur der Renaissance in Italien* (Basel, 1860), second and fourth section.

17. *The Great Art,* translated and edited by T. R. Witmer (Cambridge, Mass.: MIT Press, 1968).

Notes to Chapter 1

1. Naudé's characterization of him as untruthful is certainly the most serious slander he permitted himself (see below).

2. Henry Morley, *The Life of Girolamo Cardano of Milan, Physician,* 2 volumes (London, 1854). In addition to the *Vita propria,* a main source of information about Cardano's life and circumstances is the commentary on his own horoscope, which is given as the eighth example of the twelve genitures discussed in the appendix to the commentary on Ptolemy's *Tetrabiblos.* In the second edition of this work (*Hier. Cardani in Cl. Ptolomaei de Astrorum Iudiciis commentaria,* Basileae, 1578, mense Septembri), which was published posthumously, Cardano commented on his own life up to his sixty-eighth year (see pp. 629–680).

3. Hermann Hefele translated it into German and wrote a good introduction to it. An English translation by Jean Stoner

was published in 1930, newly edited as Dover Book (New York, 1962).

4. Jacob Burckhardt, *Die Kultur der Renaissance in Italien,* section 4, "Biographie."

See also Johann Wolfgang Goethe, *Materialien zur Geschichte der Farbenlehre,* section 3, "Hieronymus Cardanus."

5. E.g., the execution of his oldest son who had poisoned his unfaithful wife (chapter 27, 50, note by Burckhardt).

6. H. Morley, *Life of Cardano,* volume I, p. 292.

7. This is the generally accepted date of birth and the one which Cardano based his horoscope on. However, on the title page of the Basel edition of *De Subtilitate* and *De Rerum Varietate* is a circumscription of his portrait in profile which reads: "Aetatis an.XLVIIII" along with the date 1553. From this would follow that he was born in 1504 or 1505. A. Bertolotti published a last will by Cardano dated 18 January 1566, in which Cardano states his age as sixty. This would mean that he was born in 1505. (A. Bertolotti, *I Testamenti di Girolamo Cardano,* Arch. storico Lombardo, IX, p. 615, 1882). I will stay with the traditonal date without insisting on its accuracy.

8. *De Vita propria,* chapter 3.

9. For comments on this as an aspiration characteristic of the time see Jacob Burckhardt, *Kultur der Renaissance,* section 2, "The modern fame."

10. See Oystein Ore, *Cardano, the Gambling Scholar* (Princeton, N.J.: Princeton University Press, 1953).

11. In 1573, the book was translated into English at the request of Edward de Vere, 17.earl of Oxford.

12. Today we simply refer to the "cubic equation" without making a distinction as to basic forms.

13. See T. R. Witmer's introduction to *The Great Art* (Princeton, N.J.: Princeton University Press, 1968).

14. On Tartaglia, see Stillman Drake and I. E. Drabkin, *Mechanics in Sixteenth-Century Italy* (Madison, 1969).

15. Myrthe M. Cass, *The First Book of Jerome Cardan's "De Subtilitate,"* Williamsport, 1934.

16. *Opera IX,* p. 123 and p. 225.

17. *De Libris propriis, Opera I* (published in Basel in 1562 together with *Synesiorum Somniorum libri),* p. 137.

18. The crown is a gold coin approximately the size of a swiss ten-franc piece and had the purchasing power of more than one hundred Swiss francs (or fifty dollars).

19. *RV,* 88 means *De Rerum Varietate,* p. 88 of the first edition, 1557.

20. I.e., one of the four chambers forming the great court of law called "Parlement de Paris."

21. The horoscope is given as the fifth example in the aforementioned commentary on Ptolemy.

22. This edition was printed by Lodovicus Lucius.

23. I.e., grove of St. Peter. It is still a grove in the middle of a large square.

24. Jacques d'Annebaut, cardinal of Sainte Susanne. See *Dictionnaire d'Histoire et de Géographie Ecclésiastique* (Paris: Alfred Baudrillart, 1924).

25. See Edward Armstrong, *The Emperor Charles V,* 2 volumes (London, 1902).

26. Leopold v. Ranke, *Die Römischen Papste in den letzten vier Jahrhunderten,* book 3.

27. Ernest Renan, *Averroès,* 2nd edition (Paris, n.d.), p. 419.
See also Gotthold Ephraim Lessing, *Rettung des Cardano,* in the *Complete Works,* part III (Berlin, 1784).

28. Ernest Renan, *Averroès,* p. 363 ff.

29. Basel, 1554.

30. In the Basel edition of 1578 (the second edition and still prepared by Cardano) Naudé claims: "Servatoris nostri genesis non sine multorum indignatione legetur." The edition does have a special preface "ad pium lectorem," in which Cardano

defends himself against the accusation that to cast the horoscope of Christ is heresy. Inserted on page 277 is the title "Multiplicatio effectus syderum secreta, et Servatoris genesis"—but no mention is made subsequently of Christ and his geniture! The relevant passage must have been suppressed at the last moment, during the printing. In the *Opera V., Cl. Ptolemaei libri quatuor* etc., a page is inserted before chapter 10, so that page 221/222 appears twice. On the second sheet is the horoscope and its interpretation. This indicates that Spon only decided to publish the controversial passage after volume *V* had already been printed.

31. Published together with *Somniorum Synesiorum libri IV* (Basel, 1562).

32. "[H]aecceitas" comes from haecce = this! We point out a thing. Cardano thinks that the word is a scholastic barbarism, which it is. It was coined by the Scotists (Duns Scotus) and means that things can be pointed out; they are *individuals* one cannot define logically, but one can point to them.

33. J. Huizinga, *Erasmus,* translated into German by Werner Kaegi (Basel, 1951), p. 190.

Notes to Chapter 2

1. Abu Ali al Husein ibn Allah ibn Sina. Avicenna corresponds to the Spanish-Arabic pronunciation of ibn Sina. Cardano usually refers to him as the "Princeps" and emphasizes that his real name was Husein—although he writes Hasen. Sina is the name of the grandfather.

2. *Opera IX,* p. 47.

3. See Charles Singer, *A Short History of Anatomy and Physiology* (New York: Dover, 1957), p. 47 ff.

4. The spleen is an organ where blood is stored. "Black bile" is actually venous blood found in the spleen at autopsies.

5. *De Usu ciborum,* chapter 8, *Opera VII,* p. 13.

6. *De Methodo medendi,* chapter 89, *Opera VII,* p. 238. This book contains a discussion of some one hundred mistakes common in the medical practice of the time, along with suggestions for improvements. It is interesting and in character, and some of the opinions expressed are probably still valid today.

7. Ch. Singer, l.c., p. 132.

8. On the subject of rubdowns see Cardano, *De Usu ciborum,* chapter 14, *Opera VII,* p. 37.

9. A remedy to loosen phlegm.

10. Cardano regarded eiderdown as generally harmful. See *De Methodo medendi,* chapter 85 (*Opera VII,* p. 236) and *Praeceptorum ad filios liber (Opera I,* p. 475): "Super plumam non dormite!"

11. In *De Methodo medendi,* chapter 72 (*Opera VII,* p. 229) the physicians are reproached for depriving their patients of fresh air.

12. This is probably supposed to slow the ebb of venous blood, thus relieving the weakened heart-lung circulation. This seems at least to be the more recent theory, since the procedure remained in use for a long time. Cardano gives no reason for taking this measure.

13. It seems that he is afraid of catching cold while overheated.

14. Aloaesius Cornarus is the author of *Discorsi della vita sobria* (Padua, 1558). He lived to the age of about one hundred. See Jacob Burckhardt, *Die Kultur der Renaissance in Italien,* fourth section, "Biographie."

15. Since the patient is indeed suffering from states of anxiety.

16. In *De Facultatibus medicamentorum (Opera IX,* p. 367), "helleborus niger" is discussed as number 101. To Cardano, it is the "Rex medicamentorum," especially effective in purging black bile. However, if it is not properly prepared, or if it is given to a patient without preparatory treatment, it can lead to respiratory paralysis and death by suffocation.

17. As Dr. med. Felix Fierz kindly pointed out to me, the cause of the illness might have been malaria.

Notes to Chapter 3

1. *Opera I*, pp. 277–282; first printed in Basel in 1562 in a volume containing mainly *Somniorum Synesiorum libri IV*.

2. *Opera II*, pp. 283–289.

3. See J. C. Margolin, "Cardan, interprète d'Aristote," in *Platon et Aristote à la Renaissance* (Paris, 1976).

4. Festugière, *La Révélation d'Hermes Trismégiste* (Paris, 1949), volume 2, p. 370 ff.

5. Markus Fierz, "Über den Zufall," in *Spectrum Psychologiae*, Festschrift für C. A. Meier (Zurich and Stuttgart, 1965).

6. This apparently means: because the leap from the finite to the infinite is too great.

7. Ernest Renan, *Averroès et l'Averroisme* (Paris, 1861), especially chapter 3. J. C. Margolin, *Platon et Aristote à la Renaissance* (Paris, 1976).

8. Printed in 1562 in the volume containing *Somniorum Synesiorum libri, De Libris propriis*, etc.

9. It was published in a volume entitled *Opus Novum de Proportionibus Numerorum . . . in V libros digestum praeterea Artis Magnae liberimus etc. . . .* (Basel: Henri Petri, 1570). The title was apparently supplied by the publisher, since the *Liber de Proportionibus* only consists of one book, which Cardano designated as the fifth volume of his mathematical corpus.

10. T. R. Witmer, *The Great Art*, Preface, p. XVII.

11. if a_n is the age-reserve after n years, then

$$a_{n+1} = a_n - 1 - \frac{1}{40} \cdot a_n; \; a_{80} = 0.$$

12. This viewpoint was shared by David Hilbert who said: "To deny the mathematician his 'tertium non datur' would be like denying a boxer the use of his fists."

13. See page 110.

14. I think that this sentence ought to be deleted; therefore, I put it in parenthesis.

15. See page 61.

16. Nicolai de Cusa, *Trialogus de possest; Dreiergespräch über das "Können-Ist"*, ed. Renata Steiger (Hamburg, 1973), p. 55.

17. Cited after F. A. Scharf, *Nicolaus v. Cusa wichtigste Schriften.* (Freigburg i. Br., 1862), p. 123.

18. *Ibid,* p. 123.

19. One is reminded of Faust II, act I:

> . . . Um sie kein Ort, noch weniger eine Zeit . . . .
> Gestaltung, Umgestaltung
> Des ewigen Sinnes ewige Unterhaltung.
> [. . . No place, still less a time around them. . . .
> Formation, transformation
> Eternal mind's eternal entertainment.]

Since Goethe devoted a separate and very sympathetic chapter to Cardano in the *Materialien zur Geschichte der Farbenlehre,* it seems certain that he perused his works. It is well known that in such perusals, even in a mere leafing through of something, Goethe absorbed many things he found congenial.

Notes to Chapter 4

1. Pliny the Elder lived from 23 to 73 A.D. He was commander of the fleet at Misenum. The thirty-seven books of his encyclopedia, *Historia Naturalis,* were a source of knowledge for the Middle Ages and remained so well into the Renaissance. Pliny perished in the eruption of Mount Vesuvius that destroyed Herculanum and Pompeii.

2. Albertus Magnus (1193–1280) was the teacher of Thomas Aquinas. He presented contemporary scientific knowledge in the form of detailed paraphrasings of the writings of Aristotle.

3. I.e., Latin.

4. *Acts:* 17, 27–28.

5. See Aristotle, *De Insomniis.*

6. Sueton, *Octavius Augustus,* chapter 67.

7. A Jewish physician at Alexandria, 4th century. See *Physiognomika* (Basel, 1544).

8. I.e., hygroscopically dissolved potassium carbonate.

9. "[M]ens" is the Latin equivalent of the Greek "nous," whereas "spiritus" corresponds to Greek "pneuma." The "pneuma" was thought to be a very delicate material substance.

10. See *Jesus Syrach,* "Spiegel der Hauszucht," with a brief interpretation by Herm Casp. Huberinum (Nuremberg, 1580), p. 244. (This is an edition of the *Ecclesiasticus,* where each verse is followed by an edifying meditation. The author, Kaspar Huber, was probably a Lutheran pastor. The book was first published in 1552. I do not think Cardano knew this book. I only mention it because it shows that the "nine steps" was a then-current idea in quasi mystical devotion.)

11. Oystein Ore, *Cardano, the Gambling Scholar* (Princeton, N.J.: Princeton University Press, 1953).

12. These names correspond to the following scholars, who I am listing in chronological order:

Archytas of Tarent (ca. 400 B.C.), the most influential mathematician and musical theoretician of the Pythagorean school. One of his major contributions was the solution of the problem of doubling the cube. See Bartel van der Waerden, *Erwachende Wissenschaft* (Basel, Birkhäuser Verlag, 1956), p. 247.

Aristotle (384–322 B.C.)

Euclid (ca. 300 B.C. at Alexandria). His *Elements* has for two thousand years been the basic and most renowned mathematics textbook, for it not only presents geometry, but also the theory of numbers and the doctrine of irrational ratios.

Archimedes (287–212 B.C.) of Syracuse, the most significant mathematician of antiquity. His writings greatly influenced mathematicians of the seventeenth century, who venerated him as their "ancestor."

Apollonius (of Perga, ca. 200 B.C. at Alexandria). He developed the theory of conic sections: ellipse, parabola, and hyperbola.

Vitruvius. His book *De Architectura* is dedicated to Augustus. But it not only deals with architecture but also with mechanics, waterworks, sundials, etc.

Galen (129–199 A.D.), personal physician to Marcus Aurelius.

Mohammed, son of the Arab—al Khowarizmi (ca. 820 A.D.), the classical scholar of Arabic algebra. The term "algorism" was derived from his name.

Heber Hispanus—Jabir ibn Aflah (ca. 1130 at Seville), astronomer, commentator and critic of Ptolemy.

Duns (John) Scotus (died 1308), the "doctor subtilis" of scholasticism and adversary of Thomas Aquinas. Being a rigorous logician and rationalist, he emphasized that the infinite and almighty God can in no way be rationally comprehended. His writings were already being printed in the fifteenth century.

Suisset—John Swineshead (Cistercian at Oxford ca. 1350), mathematician, who in a very original manner—not geometrically but arithmetically and using infinite series—wrote the first treatise on uniformly accelerated motion.

13. Lynn Thorndyke, *History of Magic and Experimental Science,* volume VI, p. 515.

14. *Opera II,* p. 548; first printed in the volume containing *Somniorum Synesiorum libri IV* (Basel, 1562).

15. Plotinus III, 7.

16. Cardano does not use the term "imaginary space," but Manoel de Goes, the commentator of Aristotelian physics, uses it in *Commentari Collegii Conimbricensis S.J.* (1592), book 8, chapter 10, Queastio 2. This space is—as with Cardano and Parrizzi—a spiritual or divine space beyond the cosmos. In his *Experimenta Nova Magdeburgica* (Amsterdam, 1672), book 1,

chapter 35, Otto v. Guericke refers to the Conimbricensic commentary and adopts the term "imaginary space." He is, however, a consistent adherent to the Copernican world view, and consequently regards this space—as does Newton—as the divine universe.

See also:

M. Fierz, *Über den Ursprung und die Bedeutung der Lehre I.Newtons vom absoluten Raum (Gesnerus* 11 [1954]), p. 62.

———. *Das Raumproblem im 17.Jahrhundert,* in: *Connaissance scientifique et Philosophie,* Colloque organisé les 16 et 17 mai 1973 par l'Académie Royale de Belgique, p. 117 ff.

Max Jammer, *Concepts of Space* (Cambridge, Mass., 1954), p. 84 ff.

17. *Opera I,* pp. 697, 698.

18. *De Docta ignorantia,* I, p. 26.

Notes to Chapter 5

1. In the final paragraph of his *De Interrogationibus libellus* (first printed together with his *In Cl. Ptolemaei de astrorum iudiciis etc.,* second edition, Basel, 1578) Cardano says: "Never predict anything, and under no condition, to an unjust man." From this follows: "Never predict anything to an unknown person."

2. In the preface to his commentary of Ptolemy's *Tetrabiblos, Opera V.*

3. Thorndyke, op. cit. V, p. 244.

4. Margery Purver, *The Royal Society* (London, 1967), p. 31.

5. Thorndyke, op. cit., V, p. 419.

6. *Aphorismorum astronomicorum Segmenta* IV, p. 65 (*Opera V*).

7. Isaac Newton, *Opera,* ed. Samuel Horsely (London, 1785), volume V, p. 449.

8. *Aphorismorum astronomicorum Segmenta* I, p. 58 (*Opera V*).

9. *Ibid,.* V, p. 53.

10. See also the horoscope of Agrippa v. Nettesheim described above.

11. Petrarch did own Greek manuscripts, but he could not read them. See Heinrich Morf, "Die Bibliothek Petrarchas," in *Dichtung und Sprache der Romanen* (Berlin, 1922).

12. Erasmus declined the cardinalate which was offered to him.

13. *De Vita propria,* chapter 18, "Delectatio." We would like to recall here that Rabelais, himself a physician and contemporary of Cardano, was obviously inspired by Pulci's gigantic fantasies.

Notes to Chapter 6

1. *Opera I,* p. 103. Probably only the first book was written in 1545 and enlarged later on. Most of the examples in the fourth book are from a later period. Otherwise, Cardano's remark in *De Libris propriis* about how well he had compressed so much material would not make sense.

2. *Traumbuch, wahrhaftige, unbetrügliche Unterweisung wie allerhand nächtliche Träume und Erscheinungen ausgelegt werden sollen* (Basel, 1563).

3. *Heidelberger Abh. zur Philosophie und ihrer Geschichte,* edited by Wolfram Lang (Tübingen, 1926).

On Synesius of Cyrene see also H. v. Campenhausen, *Die griechischen Kirchenväter* (Vienna and Zurich, 1955).

4. It is not always necessary to recall the dream images. It may suffice to remember the feelings and emotional states.

5. This association is expressed quite beautifully in Franz Schubert's song, *Wanderer's Ode to the Moon,* which begins with the words: "I here on earth, you in the sky . . . ."

6. See Jacob Burckhardt's observations on the Vendetta in Italy in *Kultur der Renaissance,* section 6, "Morality."

7. This fine was imposed on Cardano in connection with the

trial of his son; it was illegal. The university stipend, which Cardano thought very high, amounted to seven hundred crowns.

8. That is to say, they did not repay him the one hundred crowns.

9. This does not contradict the statement that the "material" of dreams consists of memories. The question concerns the dream "content," that is, its meaning.

10. The town referred to is Metz, which in the winter of 1552/53 was besieged by Charles V and defended by Henry II of France. The siege failed and marked a turning point in the emperor's fortunes.

Notes to Chapter 7

1. This restriction was later relaxed as well, since his commentary *In libellum Hippocratis de alimento* was published in Rome in 1574.

2. *Opera I,* p. 673.

3. Carl Gustav Jung discusses this method in *Ego and Unconscious.* See also C. G. Jung, "The Transcendent Function," in *Geist und Werk,* Festschrift für Dr. Brody (Zurich, 1958).

4. The text is abridged in two places: when the father reminds Cardano of all the things he has read in Beatus, and when he talks about kings and other famous figures who spent their lives grieving.

5. *De Vita propria, Opera I,* p. 29: "It seemed to me that my soul was in the heaven of the moon, freed of the body and solitary. When I asked the meaning of this I heard my father's voice who said: I have been appointed by God as your guardian. There are souls here everywhere, but you do not see them. You will remain in this heaven for seven thousand years. You will stay that long in star after star until you have reached the eighth sphere. Then you will enter the kingdom of God."

6. Joh. Franciscus Beatus, professor in Padua about 1540, an Aristotelian. See: Jöcher's *Gelehrtenlexicon* (Leipzig, 1750/51).

7. This is not the "moon-dream." The content of that dream is indicated later on.

8. *De Vita propria, Opera I,* p. 38 ("res prorsus supra naturam").

9. Sylvester published Cardano's *De Sanitate tuenda et vita producenda* in Rome in 1580. Cardano was working on it in 1576. See *Vita propria,* chapter 45.

10. I.e., the Pope, or rather the Church and its Inquisition.

11. Adrasteia was a Thracian mother-goddess. Her name was understood by the Greeks to mean "the unavoidable," and she was, therefore, identified with Nemesis.

12. Because this would be contrary to the order of the world.

13. Tacitus, *Annalen,* 13:32.

14. Cardano was a freeman of Bologna.

Notes to Postscript

1. See Wolfgang Pauli, "Der Einfluss archetypischer Vorstellungen bei Kepler," in Carl Gustav Jung and Wolfgang Pauli, *Naturerklärung und Psyche* (Zurich, 1952).

2. In the postscript to a letter to Ph. Müller dated Linz, September 1622, he says that he is following Ptolemy and Cardano.

3. This brings to mind the "multiplication of species," a concept of medieval physics since Grossetête.

4. One must keep in mind that it is not the planets' orbits themselves that are being observed, but rather, their projection onto the celestial sphere.

5. See Alexandre Koyré, *From the Closed World to the Infinite Universe* (Baltimore, 1957).

6. See Max Jammer, *Concepts of Mass* (Cambridge, Mass., 1961).

7. The rather complicated history of developments that led to Newton's concept of force is described in R. S. Westfall, *Force in Newton's Physics* (London, 1971). On the "occultism," see B. J. T. Dobbs, *The Foundations of Newton's Alchemy* (Cambridge, 1975).

8. Christian Huygens, "Traité de la Lumière," in *Discours de la Pesenteur* (Leyden).

9. *Correspondence of Isaac Newton* (London, 1961), ed. by The Royal Society, volume 3, p. 81, note 8.

10. *Die Philosophischen Schriften Gottfried Wilhelm Leibniz,* ed. Gerhard (Berlin, 1980), volume 7, p. 337.

11. See, for example, the letter to Volder of January 19, 1706.

References

Agrippa von Nettesheim. *Die Eitelkeit und Unsicherheit der Wissenschaften.* Edited by Fritz Mauthner. 2 volumes. Munich, 1913.

Armstrong, Edward. *The Emperor Charles V.* London, 1902–13.

Bertolotti, A. *I testamenti di Girolamo Cardano.* Archivio storico Lombardo, IX. 615. 1882.

Beth, Henry. *Nicholas of Cusa.* London, 1932.

Browne, A. L. *G. Cardano's "Somniorum Synesiorum libri."* Bibl. d'hum. et Renaiss. 41 (1979).

Burckhardt, Jacob. *Die Kultur der Renaissance in Italien.* Basel, 1860.

Burtt, Edwin Arthur. *Foundations of Modern Science.* Second revised edition. 1932.

Campenhausen, Hans Freihen. *Die griechischen Kirchenväter.* Vienna and Zurich, 1955.

Cardano, Girolamo. *The Great Art.* Translated and edited by T. Richard Witmer. Boston, 1968.

Cass, Myrthe M. *The First Book of J. Cardan's "De Subtilitate."* Williamsport, 1934.

Cassirer, Ernst. *Individuum und Kosmos in der Philosophie der Renaissance.* Leipzig and Berlin, 1927.

Clagett, Marshall. *The Science of Mechanics in the Middle Ages.* Madison, 1959.

Cornaro, Aloisius. *Discorsi della vita sobria.* Padua, 1558.

Crombie, A. C. *From Augustin to Galileo.* London, 1959.

Cusa, Nicolai de. *Trialogus de possest* [Dreiergespräch über das Können-Ist]. Edited by Renata Steiger. Hamburg, 1973.

Dictionnaire d'Histoire et de Géographie Ecclésiastique. Edited by Alfred Baudrillart. Paris, 1924 ff.

Dobbs, Betty Jo Teeter. *The Foundations of Newton's Alchemy.* Cambridge, 1975.

Drake, Stillman and Drabkin, I. E. *Mechanics in Sixteenth-Century Italy.* Madison, 1969.

Festugière, O. P. *La Révélation d'Hermes Trismégiste.* 4 volumes. Paris, 1954.

Fierz, Markus, "Über den Ursprung und die Bedeutung der Lehre I. Newtons vom absoluten Raum." *Gesnerus* 11 (1954).

———. "Das Raumproblem im 17. Jahrhundert." In *Connaissance scientifique et Philosophie,* Colloque. Brussels, 1973.

———. "Über den Zufall." In *Spectrum Psychologiae,* Festschrift für C. A. Meier. Zurich and Stuttgart, 1965.

Firmiani, S. *Note e appunti su la cultura del Rinascimento. Girolamo Cardano, la vita e le opere.* Napoli, 1904.

Goethe, Johann Wolfgang v. *Materialien zur Geschichte der Farbenlehre. 3. Abt., 16. Jahrhundert,* "Cardanus."

Hilbert, David. *Die Grundlagen der Mathematik.* Hamburger Math. Einzelschriften, 5 (1928).

Huizinga, J. *Erasmus.* Translated into German by W. Kaegi. Basel, 1951.

Jammer, Max. *Concepts of Space.* Cambridge, Mass., 1954.

Jöcher, Christian Gottl. *Allgemeines Gelehrtenlexicon.* 4 volumes. Leipzig, 1750–51.

Jung, Carl Gustav. *Die Beziehung zwischen dem Ich und dem Unbewussten.* Zurich, 1933.

__________. "Die transcendente Funktion." In *Festschrift für D. Brody.* Zurich, 1958.

Kepler, Johannes. *Briefe.* Edited by Max Caspar and Walther von Dyk. Munich, 1930.

Koyré, Alexandre. *From the Closed World to the Infinite Universe.* Baltimore, 1957.

Lang, Wolfram. *Heidelberger Abhandlungen zur Philosophie und ihrer Geschichte.* Tübingen, 1926.

Lasswitz, Kurd. *Geschichte des Atomismus.,* 2 volumes. Hamburg, 1890.

Leibniz, Gottfried Wilhelm. *Philosophische Schriften.* Edited by Gerhard. Volume 7. Berlin, 1890.

Lessing, Gotthold Ephraim. *Sämmtliche Schriften.* Part 3, "Rettung des Cardano." Berlin, 1784.

Margolin, J. C. "Cardan interprète d'Aristote." In *Platon et Aristote à la Renaissance.* Volume 32 of the series *De Petrarque à Descartes.* Edited by P. Mesnard. Paris, 1976.

Morf, Heinrich. "Die Bibliothek Petrarcas." In *Sprache und Dichtung der Romanen.* Berlin, 1922.

Morley, Henry. *The Life of Girolamo Cardano.* 2 volumes. London, 1854.

Newton, Isaac. *Opera.* Edited by Samuel Horsely. Volume 5. London, 1785.

Ore, Oystein. *Cardan, the Gambling Scholar.* Princeton, N.J.: Princeton University Press, 1953.

Patritius, Franciscus. *Mova de Universis Philosophia.* Venice, 1593.

Pauli, Wolfgang. "Der Einfluss archetypischer Vorstellungen bei Kepler." In Carl Gustav Jung und W. Pauli, *Naturerklärung und Psyche.* Zurich, 1952.

Purver, Margery. *The Royal Society: Concept and Creation.* London, 1967.

Ranke, Leopold v. *Die römischen Päpste, ihre Kirche und ihr Staat im 16. und 17. Jahrhundert* (1834–36). Sixth edition, 1874.

__________. *Deutsche Geschichte im Zeitalter der Reformation.* Fifth edition, 1873.

Regiomontanus, Johannes. *De Triangulis libri V accessit Cusani de Quadratura circuli.* Norimbergae, 1533.

Renan, Ernest. *Averroès et l'Averroisme.* Paris, 1861. Especially chapter 3, "l'Averroisme dans l'école de Padone."

Rixner, Thaddae and Siber, Thaddae. *Leben und Lehrmeinungen berühmter Physiker am Ende des 16. Jahrhunderts.* Book 2: "Cardano." Sulzbach, 1820.

Russell, Bertrand. *A History of Western Philosophy.* New York, 1945.

Scharff, F. A. *Nicolaus von Cusa wichtigste Schriften.* Freiburg i. Br., 1862.

Singer, Charles. *A Short History of Anatomy and Physiology.* New York, 1957.

Thorndyke, L. *History of Magic and Experimental Science.* 8 volumes. New York, 1923–58.

Van der Waerden, Bartel. *Erwachende Wissenschaft.* Basel, 1956.

Westfall, Richard S. *Force in Newton's Physics.* London, 1971.

Appendix

Early Translations

De Subtilitate

De la Subtilité, Traduit par Rich. LeBlanc à Paris, chez le Noir, 1556.

De Rerum Varietate

Offenbarung der Natur . . . auch mancherlei subtiler Würkungen . . . durch Heinrich Pantaleon verdeutscht. Basel, 1559.

Offenbarung der Natur und natürlicher Dinge, aus dem Latein von Hulrich Fröhlich. Basel, 1591.

Somniorum Synesiorum libri

Traumbuch Cardani . . . , verdeutscht durch Joh. Jak. Huggelinum. Basel, 1563.

De Consolatione

Cardanus Comforte, translated into Englishe and published by commandement of the right Honourable the Earl of Oxforde. 1573.

Metoposkopie (français). Paris, 1658.

Proxeneta seu prudentia civilis

Science du Monde ou Sagesse civile. 4me ed., 12°. Paris, 1661.

Modern Translations

Vita propria

Vicenzio Mantovani, *Vita per lu medisimo.* 1821.

Hermann Hefele, *Des Girolamo Cardano eigene Lebensbeschreibung.* Jena, 1914.

Jean Stoner, *The Book of My Life.* New York, 1930.

Jean Dayre, Edition et traduction. Paris, 1936.

Ars Magna

T. R. Witmer, Translator and editor, with a forward by Oystein Ore. Cambridge, Mass., 1968.

De ludo aleis

Oystein Ore, *Cardano, the Gambling Scholar.* With translation "Book on Games," by Sydney Hen. Gould. Princeton, 1953.

De Subtilitate

Myrthe Marguerite Cass, *The First Book of Jer. Cardan's "de Subtilitate."* Williamsport, 1934. [Includes a list of probably all editions of *De Subtilitate.*]

Robert Hooke owned the following works by Cardano:

De Sapientia et de consolatione (1544 and 1624)
Vita propria (1654)
Somniorum Synesiorum libri IV etc. (1562, ed. princeps)
Prudentia civilis (1635)
De Utilitate ex adversis capienda (1672)
De Venenis (1653)
De Subtilitate (1582). (Basel: Henric Petri, third ed.)
De Rerum Varietate (1553, ed. princeps)
De Proportionibus et Ars magna etc. (1570, ed. princeps)
(See: *Sale Catalogues of Libraries of eminent persons,* edited by Feinberger, volume 11, "Scientists." London, 1975.

John Locke owned:

Opera omnia, 10 volumes fol. (1663)
De Rerum Varietate, 8 (Basel, 1581)
De Utilitate ex adversis capienda (Basel 1561, ed. princeps)
Proxeneta seu prudentia civili (1627, ed. princeps, and 1635)
See: J. Harrison and P. Laslett, *The Library of John Locke,* second ed. (1971).

First editions of the works of Cardano, by himself prepared for publication

1. *Pronostico o vero judicio* . . . . 1534–50. 8° 11 fol. (Venetiis, 1534). (P)
2. *De malo recentiorum medendi usu libellus + libellus de simpliciorum medicinarum noxa.* 8° (Venetiis, 1536). (L)
3. *Libellus qui dicitur supplementum in almanach + Libellus de restitutione temporum + quinque principum geniturae cum expositione + quinque eruditum virorum geniturae cum exposione.* 4° (Mediolani, 1538). (L)
4. *Practica arithmetica.* (Mediolani, 1539). (P, L)
5. *De consolatione libri tres.* 8° 132 fol. (Venetiis, 1542). (P, L)
6. *De supplemento in almanach item geniturae 67 insignis casibus.* 4° (Norimbergae, 1543). (P, B, L)
7. *De sapientia + De consolatione + De libris propriis.* 4° 341 pp. (Norimbergae, 1544). (P, L)
8. *De immortalitate animorum.* (Lyons, 1545. 1st ed. ?). (P)

9. *Contradicentium medicorum liber.* 8° 188 fol. (Venetiis, 1545). (N)
10. *Ars magna sive de regulis algebraicis.* fol. 81 pp. (Norimbergae, 1545). (P, B, L)
11. *De supplemento in almanach + de iudiciis geniturarum + de revolutionibus + de exemplis centum geniturarum + aphorismorum astronomicorum liber.* 4° 310 pp. (Norimbergae, 1547). (P, B, L)
12. *De subtilitate libri XXI.* fol. 390 pp. (Norimbergae, 1550). (P, B, L)
13. *De subtilitate libri XXI,* nunc demum recogniti atque perfecti. fol. 561 pp. (Basileae, 1554). (P, B, L)
14. *In Cl. Ptolemaei Pelusiensis IIII de astrorum iudiciis commentaria praeterea geniturarum XII exempla.* fol. 513 pp. (Basileae, 1554). (P, B, L)
15. *De rerum varietate libri XVII.* fol. 707 pp. 8° 2 volumes. (Basileae, 1557). (P, B)
16. *Opuscula: de aqua et aethere + de Cyna radica + consilium pro H. Palavicino + consilium pro fluxa sanguinis coercendo + consilium pro Mantuano lepram patiente + medicine encomium + in calumniatorem librorum de subtilitate actio.* fol. (Basileae, 1559). (B, L)
17. *De subtilitate libri XXI ab authore plusquam mille locis illustrati. Addita insuper apologia adversus calumniatorem, quavis horum librorum aperitur.* fol., 8° (Basileae, 1560). (P, B, L)
18. *De utilitate ex adversis capienda + defensiones pro filio coram praeside provinciae et senatu habitae +* Jo. Babt. *Cardani de abstinentia ab usu ciborum foetidorum libellus exiguus quem moriens explere non potuit.* 8° 1161 pp. (Basileae, 1561). (P, B)

19. *Somniorum Synesiorum libri IV + De libris propriis + de curationibus et praedictionibus + Neronis encomium + De uno + actio in Thessalonicum medicum + de secretis + de gemmis et coloribus + dialogus de morte et de humanis consiliis Tetim inscriptus + De minimis et propinquis + De summo bono.* 4° 775 pp. (Basileae, 1562). (P, B, L)

20. *In septem aphorismorum Hippocratis particulas commentaria + De venenorum differentiis libri tres + De providentia temporum liber.* fol. 536 pp. (Basileae, 1564). (P, B, L)

21. *Ars curandi parva et alia, nunc primum aedita, opera in duos tomos divisa.*

a) *Ars curandi parva + consilium datum Rev. D.D. Joan + ephemeris Rev. D.D. Joan Amulton + consilium pro D. Hier. Palavicino Marchione + Ventriculi dolore ob morbum gallicum laboranti mercatoré Mediolanensi consilium Hippocraticum + ad surditatem consilium pro Francisco cive Genuensi + sanguinis fluxam coercendi consilium generale + pro lepra patiente consilium + medicinae encomium + podagrae encomium + apologia ad And. Canutium.* 8°, 720 pp.

b) *Dialectica + Hyperchen + de Socrate studio + Antigorgias dialogus seu de recta vita ratione + de aqua liber et de aethere liber + de Cyna radice.* 8°, 621 pp. (Basileae, 1566). (P, B, L)

22. *In Hippocratis . . . Prognostica, atque in Galeni prognosticorum expositione commentarii + in libros Hippocratis de septimestri et octomestri partu et simul in eorum Galeni commentaria Cardani commentaria + consilia septem.* 4°, 831 pp. (Basileae, 1568). (P, B, L)

23. *Commentarii in Hippocratis de aere, aquis et locis, opus*

divinum + Hier. Cardani ad Aliciatum oratio quam triplici Geryonis canem appellat item Joa. Bapt. Cardani de fulgure liber unus item Hier. Cardani consilia tria. fol. 338 pp. (Basileae, 1570). (P, B, L)

24. *De proportionibus liber + Ars magna + de Aliza regula.* fol. 545 pp. (Basileae, 1570). (P, B, L)
25. *In libellum Hippocratis de alimento commentaria.* 8°, 250 pp. (Romae, 1574). (P)
26. *De septem erraticarum stellarum qualitatibus atque viribus + in Ptolomaei de astrorum iudicia commentaria + geniturarum exempla XII + de rogationibus libellum.* fol. 510 pp. + 115 pp. (Basileae, 1578). (P, B, L)
27. *De sanitate tuenda et vita producenda,* a Rud. Sylvestrium editum. fol. 332 pp. (Romae, 1580). (P)
28. *De subtilitate libri XXI ab authore plus/quam mille locis illustrati, additur insuper apologia adversus calumniatorem etc.* fol. (Basileae, 1582). (B, L)
29. *De causis signis et locis.* 8°, 244 pp. (Basileae, 1583). (P, B)

Published posthumously:

30. *Proxeneta seu de prudentia civili.* 12°, 767 pp. (Lugdunum Batavorum, 1627).
31. *Praecepta ad filios liber,* ed. Naudé. (Paris, 1635).
32. *De propria vita,* ed. Naudé. (Paris, 1648).
33. *Metoposkopia.* fol. 227 pp. (Paris, 1658).

N.B.: P: Paris, Bibliothèque Nationale
L: London, British Library
B: Basel, Universitäts- Bibliothek
N: National Union Catalogue, volume 95 (1970), U.S.A.

Cf.: *Vita propria,* chapter 45.